Spon's First Stage Estimating Handbook

Second edition

Other books by Bryan Spain

also available from Taylor & Francis

Spon's Estimating Costs Guide to Plumbing & Heating
(2006 edition)
Pb: 0-415-38618-7

Spon's Estimating Costs Guide to Electrical Works (2006 edition)
Pb: 0-415-38614-4

Spon's Estimating Costs Guide to Minor Works, Alterations and Repairs to Fire, Flood, Gale and Theft Damage (2006 edition)
Pb: 0-415-38213-0

Spon's Estimating Costs Guide to Roofing (2005 edition)
Pb: 0-415-34412-3

Spon's Estimating Costs Guide to Finishings: painting and decorating, plastering and tiling (2005 edition)
Pb: 0-415-34411-5

Spon's Estimating Costs Guide to Minor Landscaping, Gardening and External Works
(2005 edition)
Pb: 0-415-34410-7

Spon's House Improvement Price Book: house extensions, storm damage work, alterations, loft conversions and insulation
(2005 edition)
Pb: 0-415-37043-4

Spon's Construction Resource Handbook (1998 edition)
Hb: 0-419-23680-5

Information and ordering details

For price availability and ordering visit our website
www.tandfbuiltenvironment.com

Alternatively our books are available from all good bookshops.

Spon's First Stage Estimating Handbook

Second edition

Bryan Spain

Taylor & Francis
Taylor & Francis Group

LONDON AND NEW YORK

First published 2000 by Spon Press
Second edition published 2006 by Taylor & Francis
2 Park Square, Milton Park, Abingdon, Oxon OX14 4RN

Simultaneously published in the USA and Canada
by Taylor & Francis
270 Madison Avenue, New York, NY 10016, USA

Reprinted 2008

Taylor & Francis is an imprint of the Taylor & Francis Group, an informa business

Publisher's note
This book has been prepared from camera-ready copy supplied by the author

Printed and bound in Great Britain by TJ International Ltd, Padstow, Cornwall

British Library Cataloguing in Publication Data
A catalogue record for this book is available from the British Library

ISBN 10: 0-415-38619-5 (pbk)
ISBN 10: 0-203-96973-1 (ebk)

ISBN 13: 978-0-415-38619-7 (pbk)
ISBN 13: 978-0-203-96973-1 (ebk)

Contents

Preface

Assessing the total cost of a construction project is a continuous process that commences when the client asks his professional advisors what it will cost. It ends when the cheque for the release of the last portion of the retention is paid and the final account is settled.

Whether a project is profitable or not can often depend on decisions taken at the first stages of its life. The people who carry out this assessment bear the responsibility of deciding whether the cost information available at the time warrants spending further time and money investigating the viability of a project that is usually still at the conceptual stage.

It is vital, therefore, that the most accurate methods of calculating the probable costs of construction are used and I hope that the information in this book will help clients, developers, architects, engineers and surveyors involved in this process.

Time spent on the appraisal of projects that do not proceed is not always wasted – lessons can be learnt that will help the next project. The greater waste lies in allowing jobs to proceed without the people making the key decisions having the best cost information available to them.

It is not possible to provide totally accurate cost data in the early stages of a job – only a historical analysis of the final account can do that. But guidelines can be set out that should allow the client commissioning the work to help make the right decision before becoming contractually committed to spending large sums of money on land purchase, professional fees and construction costs.

On a well-managed project, the cost plan will be monitored continuously as changes are made to the original design or for other reasons. If the first-stage budget is based on reliable cost data, the project will have a better chance of being completed within budget.

I have received a great deal of help in the preparation of this book and would like to thank those suppliers and contractors who gave me their time and support. The information in Estimating Data on civil engineering outputs is based on data in Spon's Civil Engineering and Highways Price Book 2006 edited by Davis Langdon.

Although every effort has been made to ensure the accuracy of the information, neither the Publishers nor I can accept any type of liability resulting from the use of the contents.

On a more positive note, I would welcome constructive criticism of ways to make the contents more relevant to the changing needs of the construction industry for future editions.

Bryan Spain
January 2006

Introduction

This second edition of Spon's First Stage Estimating Handbook is aimed at all members of the construction team who are involved in assessing construction costs during the early stages of the development of a project. There are many factors that can affect the profitability of a job over which the construction team has little or no control.

It is important, therefore, that areas that can be influenced by good management skills should be given special attention. This applies particularly to the first stage estimating process when a solid base of the project's financial position can be established.

The contents have been laid out in the order that most projects are financially assessed. The quality of the cost information, however, is almost wholly dependent upon the quality of the specification and design information available at that stage of the appraisal.

Chapter 1 lists square metre rates for building and M & E work for a wide variety of building types. The rates are expressed in a range based on historical costs and should provide a client with a broad indication of likely costs.

Chapter 2 fulfils the same function but the costs are given in unit costs, e.g. the cost per cinema seat or per cover in a restaurant. This method of assessment can be surprisingly accurate but should always be backed up by more detailed methods of cost appraisal.

Chapter 3 contains elemental cost analyses of 32 different types of buildings. These analyses provide data on the percentage and cost breakdown of 24 elements for each building and are a useful tool in identifying imbalances between the elements of different buildings.

Chapter 4 includes unit rates for building, landscaping, civil engineering, M&E and alteration work. Where possible, composite rates are displayed that combine several different descriptions to provide a single rate for separate but linked activities. For example, excavation, concrete and brickwork up to DPC level are combined to produce a single linear metre rate for strip foundations.

The use of composite rates can save valuable time in the preparation of the cost plan. Where item descriptions do not lend themselves to be combined with others they are listed as principal rates.

Chapter 5 provides indices reflecting historical costs of construction costs and tender prices.

Chapter 6 deals with property insurance and contains an example of re-building costs for insurance purposes.

Chapter 7 sets out the cost of employing professional advisors whose fees are now negotiable and not mandatory.

Chapter 8 contains information on life cycle costing which is now recognised as an important factor in the assessment of overall construction costs.

Chapter 9 includes lists of useful addresses.

Chapter 10 contains information that may be useful when preparing a first stage estimate.

At this stage of a project the information available is limited and, although the brief from a client can be minimal, an approximation of costs must still be prepared. Here is a simple example.

It is proposed to construct a single story distribution warehouse (shell only) size 6,000 square metres in the South West of England. An outline plan has been prepared showing parking areas and a road layout. It is known that local contractors are operating at full capacity so tender levels are expected to be high. The work would commence in April 2006 with a nine month construction period.

Main building		£
6,000 m2 at £500 (Chapter 1)		3,000,000

External works

(based on rough quantities and rates in Chapter 4)

Car park		180,000	
Roads		135,000	
Fencing		46,000	
Drains		42,000	
Soft landscaping		15,000	
Lighting		15,000	433,000
	£		3,433,000

Extra for anticipated high tender levels		
say 6%		205,980
	£	3,638,980
Allow for inflation		
say 9 months @ 3% x £906,300		81,877
Carried forward	£	3,720,857

Brought forward		£3,720,857

Allow for regional variations (see Chapter 5)
say South West, 86% x £3,720,857

Professional fees (see chapter 5)

say 12% x 3,199,937	£	<u>383,992</u>
	£	<u>3,583,929</u>

Depending on the quality of the information on which the first stage estimate is based, a basement and ceiling band of between 5% and 10% should be applied and reported to the client:

5%	say	£3,405,000 to £3,763,000
10%	say	£3,226,000 to £3,942,000

These costs exclude land purchase and VAT and any special costs connected to the project. They also exclude any development grants or subsidies that may be available in the area involved.

Costs per square metre

BUILDING WORK

These square metre prices exclude the cost of external works, fittings, furniture, professional fees and VAT. They are expressed in a range and intended to provide a broad indication of the cost of the work.

Public service buildings	£/m2
Banks	
local	1400-1800
city	1750-2250
Building societies	
local	1300-1700
city	1600-2100
Fire stations	1200-1650
Courts	
magistrates	1450-1700
county	1650-1900
Police stations	1150-1500
Prisons	1600-1850
Post offices	800-1200
Halls	
town	950-1250
village	900-1200
Industrial buildings	
Agricultural	
livestock	450-600
storage	400-550

2 Costs per square metre

	£/m2
Factories	
light industrial	500-650
heavy industrial	650-900
extra for owner occupation	200-250
extra for office accommodation	250-300
High tech. laboratories	2500-3000
extra for air conditioning	150-200
Warehouses/stores shell only	
up to 2000m2 floor area	450-550
over 2000m2 floor area	350-500
extra for owner occupation	200-250
Offices, business park	
high tech unit	1000-1300
extra for air conditioning	150-200
extra for owner occupation	200-250
Offices, city centre	
3 to 5 floors	1100-1400
over 5 floors	1250-1650
extra for air conditioning	150-200
prestige building	2000-2800
extra for owner occupation	200-250

Health and welfare facilities

Day surgeries	900-1250
Group surgeries	950-1300
Homes for the mentally handicapped	1000-1400
Health centres	1200-1550

£/m2

Hospitals
general	1300-1700
laboratories	1550-1950
pharmacies	1000-1500
private	1400-1800
teaching centres	980-1400

Nursing homes   900-1250

Leisure

Cinemas
shell only	600-1050
complete	950-1300

Community centres   900-1250

Concert halls   1800-2500

Exhibition buildings   1250-1650

Ice rinks   1250-1550

Golf club houses   1300-1800

Public houses   1100-1350

Restaurants
shell only	600-850
complete	1100-1450

Sports centres   700-1050

Sports pavilions   1100-1400

Squash courts   850-1200

4 Costs per square metre

	£/m2
Swimming pools	
school	1100-1300
local	1300-1600
international	1600-2500
fun	1800-2200
Theatres	
small	1450-1750
prestige	1650-2200

Religious buildings

Churches, chapels	1100-1400
Church halls, meeting houses	1150-1350
Convents	1200-1350
Crematoria	1450-1750
Temples, synagogues, mosques	1250-1700

Educational buildings

Colleges	1100-1450
Laboratories	1750-2250
Libraries	
local	950-1250
city centre	1350-1650
Museums	
local	1500-1900
city centre	1650-2100

	£/m2
Research facilities	1600-2000

Schools

nursery	1200-1500
primary	1100-1400
secondary	1050-1350
special	1050-1350

Universities	1250-1600

Residential buildings

Hotel

3 star	1400-1750
5 star	2000-2500

Private housing

bungalows	800-950
houses, detached	700-900
houses, semi-detached	700-850
flats, low rise	700-800
flats, standard	700-800
flats, luxury	850-1150

Public housing

bungalows	850-950
houses, semi-detached	700-900
flats, low rise	750-950
sheltered housing, one storey	750-950
sheltered housing, two storey	700-900

Student accommodation	1100-1350

Youth hostels	800-1150

Transport

Aircraft hangars	1500-2500

Airport terminals	1600-2700

Air traffic control buildings	1550-1900

6 Costs per square metre

	£/m2
Car parking	
multi-storey	300-500
underground	500-600
surface	75-100
Car show rooms	750-1000
Coach and bus stations	900-1300

Retail

	£/m2
Fast food units	
shell only	500-750
complete	1100-1400
Retail warehouses	
shell only	400-550
complete	600-800
Shops	
shell only	450-750
complete	1150-1350
Shopping malls	
shell only	450-800
complete	2500-3000
Supermarkets	
shell only	400-750
complete	1200-1650

MECHANICAL AND ELECTRICAL WORK

These rates per square metre are for all engineering systems of specific building types applied to the gross internal floor area of buildings excluding professional fees and VAT. They are expressed in a range and are intended to provide a broad indication of the cost of the work.

Public service buildings	£/m2
Ambulance stations	160-200
Banks	350-310
Council offices	250-300
Fire stations	300-330
Magistrates courts	230-290
Police stations	220-260
Industrial buildings	
Agricultural sheds	90-120
Factories	
advance	90-120
high tech.	180-220
purpose built	140-180
Warehouses	90-120
Workshops (excluding machinery)	100-130
Educational buildings	
Laboratories	230-300

8 Costs per square metre

	£/m2
Primary schools	180-250
Secondary schools	160-200
Universities	170-230
Leisure	
Community centres	240-290
Gymnasia	160-190
Hotels	400-500
Public houses	200-270
Restaurants	230-290
Social centres	240-290
Sports halls	130-170
Swimming pools/sports facilities	180-240
Health and welfare facilities	
Childrens' homes	290-360
General hospitals	400-450
Health centres	190-250

£/m2

Residential buildings

Local Authority and Housing Association schemes

high rise flats	150-190
4/5 person, two storey houses	130-190

2

Costs per unit

The following represent the cost per unit of a range of types of buildings. This method of assessing construction costs is usually the first step in the process of the consideration of the viability of project.

	Unit	£
Health and welfare facilities		
Hospitals (general)	Bed	90,000-130,000
Hospitals (private)	Room	115,000-200,000
Nursing home		
old peoples	Bed	30,000-45,000
childrens	Bed	25,000-40,000
Leisure		
Theatres	Seat	20,000-30,000
Sports stadium, new stands		
single tier stand	Seat	750-1,250
multi-tier stands with		
hospitality boxes	Seat	1,000-1,500
Hotels		
3 star	Bedroom	15,000-25,000
5 star	Bedroom	30,000-90,000
Residential		
Local Authority housing	Bedroom	9,000-11,000
Local Authority flats	Bedroom	14,000-16,000

12   Costs per unit

	Unit	£
Sheltered housing	Bedroom	16,000-22,000
Private housing	Bedroom	20,000-25,000
Private flats	Bedroom	25,000-35,000
Student accommodation	Bedroom	15,000-20,000

Transport

Car parking		
multi storey	Car	5,000-10,000
underground	Car	15,000-25,000
surface	Car	1,200-1,750

Educational buildings

Nursery schools	Pupil	5,000-10,000
Primary schools	Pupil	7,000-12,000
Secondary schools	Pupil	8,000-14,000

3

Elemental costs

This chapter contains elemental cost breakdowns for building, mechanical and electrical work. The building work is comprised of 33 different types in 8 categories, each building broken down into 24 cost elements. The M & E section contains elemental costs of 6 types of buildings.

The cost data should be used with caution because imbalances can occur in the use of elemental costs in isolation. The figures are an amalgam of the costs for a range of buildings in each category and represent a broad indication of costs rather than an accurate statement of detail.

Nevertheless, the information provided should be an invaluable tool in assessing the relative values of elements in different buildings. The tables will be particularly useful in the early cost planning process and also in the evaluation of tenders.

The costs cover building work only and exclude the costs of drainage, external works, contingency sums, professional fees and VAT, Due to rounding off there may be some minor discrepancies in individual elemental costs and totals. The figures are based on costs prevailing in the first quarter of 2006. The following buildings are included.

BUILDING WORK

Public service buildings

Ambulance station
City bank
County court
Fire station
Police station
Village hall

Industrial buildings

Factory, light industrial
Factory, heavy industrial
Livestock building
High tech laboratory
Nursery units

Industrial buildings (cont'd)

Warehouse shell
Warehouse complete

Health and welfare

Group surgery
Health centre
Old persons' nursing home
Welfare centre

Leisure

Golf club house
Public house
Restaurant
Sports hall

Education

Library
Primary school
Secondary school
Sixth form college
Special school
Teachers' training college

Residential

Local Authority low rise flats
Local Authority housing
Private flats
Luxury private flats

Transport

Multi-storey car park

MECHANICAL AND ELECTRICAL WORK

Hotel
Hospital
Leisure complex
Office block
Sports stadium
University

Category: Public service
Type: Ambulance station
Floor area: 375m2
Total cost: £ 341,250

	Elemental cost £	% of cost	Cost per m2 £
1 Preliminaries	37,538	11	100
2 Substructure	30,713	9	82
3 Frame	0	0	0
4 Upper floors	0	0	0
5 Roof	44,363	13	118
6 Staircases	0	0	0
7 External walls	27,300	8	73
8 Windows and external doors	13,650	4	36
9 Partitions and internal walls	10,238	3	27
10 Internal doors	6,825	2	18
11 Wall finishes	6,825	2	18
12 Floor finishes	17,063	5	46
13 Ceiling finishes	3,413	1	9
14 Fittings and furnishings	23,888	7	64
15 Sanitary appliances/disposal installation	13,650	4	36
16 Hot and cold water services	6,825	2	18
17 Heating and air treatment installation	13,650	4	36
18 Ventilation installation	3,413	1	9
19 Gas services	3,413	1	9
20 Electric installation	27,300	8	73
21 Lift and conveyor installation	0	0	0
22 Protective & communication installation	6,825	2	18
23 Special installations/services equipment	13,650	4	36
24 Builders' work	30,713	9	82
	341,250	100	910

Category: Public service
Type: City bank
Floor area: 907m2
Total cost: £1,337,825

	Elemental cost £	% of cost	Cost per m2 £
1 Preliminaries	120,404	9	133
2 Substructure	93,648	7	103
3 Frame	133,783	10	148
4 Upper floors	53,513	4	59
5 Roof	80,270	6	89
6 Staircases	26,757	2	30
7 External walls	200,674	15	221
8 Windows and external doors	66,891	5	74
9 Partitions and internal walls	40,135	3	44
10 Internal doors	26,757	2	30
11 Wall finishes	26,757	2	30
12 Floor finishes	53,513	4	59
13 Ceiling finishes	40,135	3	44
14 Fittings and furnishings	40,135	3	44
15 Sanitary appliances/disposal installation	13,378	1	15
16 Hot and cold water services	26,757	2	30
17 Heating and air treatment installation	40,135	3	44
18 Ventilation installation	13,378	1	15
19 Gas services	0	0	0
20 Electric installation	66,891	5	74
21 Lift and conveyor installation	0	0	0
22 Protective & communication installation	40,135	3	44
23 Special installations/services equipment	40,135	3	44
24 Builders' work	93,648	7	103
	1,337,825	100	1,475

Category: Public service
Type: County court
Floor area: 948m2
Total cost: £1,311,084

	Elemental cost £	% of cost	Cost per m2 £
1 Preliminaries	131,108	10	138
2 Substructure	65,554	5	69
3 Frame	91,776	7	97
4 Upper floors	65,554	5	69
5 Roof	91,776	7	97
6 Staircases	39,333	3	41
7 External walls	117,998	9	124
8 Windows and external doors	65,554	5	69
9 Partitions and internal walls	39,333	3	41
10 Internal doors	52,443	4	55
11 Wall finishes	39,333	3	41
12 Floor finishes	52,443	4	55
13 Ceiling finishes	39,333	3	41
14 Fittings and furnishings	65,554	5	69
15 Sanitary appliances/disposal installation	26,222	2	28
16 Hot and cold water services	13,111	1	14
17 Heating and air treatment installation	26,222	2	28
18 Ventilation installation	13,111	1	14
19 Gas services	13,111	1	14
20 Electric installation	78,665	6	83
21 Lift and conveyor installation	26,222	2	28
22 Protective & communication installation	13,111	1	14
23 Special installations/services equipment	26,222	2	28
24 Builders' work	117,998	9	124
	1,311,084	100	1,383

Category: Public service
Type: Fire station
Floor area: 490m2
Total cost: £650,720

	Elemental cost £	% of cost	Cost per m2 £
1 Preliminaries	65,072	10	133
2 Substructure	39,043	6	80
3 Frame	488,040	5	996
4 Upper floors	19,522	3	40
5 Roof	52,058	8	106
6 Staircases	13,014	2	27
7 External walls	65,072	10	133
8 Windows and external doors	39,043	6	80
9 Partitions and internal walls	19,522	3	40
10 Internal doors	19,522	3	40
11 Wall finishes	19,522	3	40
12 Floor finishes	26,029	4	53
13 Ceiling finishes	6,507	1	13
14 Fittings and furnishings	13,014	2	27
15 Sanitary appliances/disposal installation	13,014	2	27
16 Hot and cold water services	19,522	3	40
17 Heating and air treatment installation	19,522	3	40
18 Ventilation installation	6,507	1	13
19 Gas services	6,507	1	13
20 Electric installation	65,072	10	133
21 Lift and conveyor installation	0	0	0
22 Protective & communication installation	13,014	2	27
23 Special installations/services equipment	19,522	3	40
24 Builders' work	58,565	9	120
	650,720	100	1,328

Category: Public service
Type: Police station
Floor area: 342m2
Total cost: £434,682

	Elemental cost £	% of cost	Cost per m2 £
1 Preliminaries	47,815	11	140
2 Substructure	30,428	7	89
3 Frame	21,734	5	64
4 Upper floors	8,694	2	25
5 Roof	30,428	7	89
6 Staircases	13,040	3	38
7 External walls	47,815	11	140
8 Windows and external doors	30,428	7	89
9 Partitions and internal walls	17,387	4	51
10 Internal doors	13,040	3	38
11 Wall finishes	13,040	3	38
12 Floor finishes	17,387	4	51
13 Ceiling finishes	8,694	2	25
14 Fittings and furnishings	8,694	2	25
15 Sanitary appliances/disposal installation	8,694	2	25
16 Hot and cold water services	8,694	2	25
17 Heating and air treatment installation	8,694	2	25
18 Ventilation installation	4,347	1	13
19 Gas services	4,347	1	13
20 Electric installation	34,775	8	102
21 Lift and conveyor installation	0	0	0
22 Protective & communication installation	13,040	3	38
23 Special installations/services equipment	8,694	2	25
24 Builders' work	34,775	8	102
	434,682	100	1,271

Category: Public service
Type: Village hall
Floor area: 380m2
Total cost: £300,580

	Elemental cost £	% of cost	Cost per m2 £
1 Preliminaries	33,064	11	87
2 Substructure	30,058	10	79
3 Frame	0	0	0
4 Upper floors	0	0	0
5 Roof	27,052	9	71
6 Staircases	0	0	0
7 External walls	30,058	10	79
8 Windows and external doors	21,041	7	55
9 Partitions and internal walls	12,023	4	32
10 Internal doors	9,017	3	24
11 Wall finishes	9,017	3	24
12 Floor finishes	12,023	4	32
13 Ceiling finishes	6,012	2	16
14 Fittings and furnishings	15,029	5	40
15 Sanitary appliances/disposal installation	6,012	2	16
16 Hot and cold water services	3,006	1	8
17 Heating and air treatment installation	9,017	3	24
18 Ventilation installation	0	0	0
19 Gas services	0	0	0
20 Electric installation	36,070	12	95
21 Lift and conveyor installation	0	0	0
22 Protective & communication installation	9,017	3	24
23 Special installations/services equipment	3,006	1	8
24 Builders' work	30,058	10	79
	300,580	100	791

Category: Industrial
Type: Factory, light industrial
Floor area: 380m2
Total cost: £2,211,780

	Elemental cost £	% of cost	Cost per m2 £
1 Preliminaries	154,825	7	41
2 Substructure	287,531	13	75
3 Frame	353,885	16	93
4 Upper floors	0	0	0
5 Roof	221,178	10	58
6 Staircases	0	0	0
7 External walls	221,178	10	58
8 Windows and external doors	88,471	4	23
9 Partitions and internal walls	66,353	3	17
10 Internal doors	22,118	1	6
11 Wall finishes	22,118	1	6
12 Floor finishes	44,236	2	12
13 Ceiling finishes	22,118	1	6
14 Fittings and furnishings	176,942	8	46
15 Sanitary appliances/disposal installation	22,118	1	6
16 Hot and cold water services	22,118	1	6
17 Heating and air treatment installation	88,471	4	23
18 Ventilation installation	22,118	1	6
19 Gas services	22,118	1	6
20 Electric installation	176,942	8	46
21 Lift and conveyor installation	0	0	0
22 Protective & communication installation	132,707	6	35
23 Special installations/services equipment	22,118	1	6
24 Builders' work	22,118	1	6
	2,211,780	100	579

Category: Industrial
Type: Factory, heavy industrial
Floor area: 10,500m2
Total cost: £6,373,500

	Elemental cost £	% of cost	Cost per m2 £
1 Preliminaries	573,615	9	55
2 Substructure	637,350	10	61
3 Frame	764,820	12	73
4 Upper floors	254,940	4	24
5 Roof	573,615	9	55
6 Staircases	63,735	1	6
7 External walls	509,880	8	49
8 Windows and external doors	191,205	3	18
9 Partitions and internal walls	127,470	2	12
10 Internal doors	63,735	1	6
11 Wall finishes	127,470	2	12
12 Floor finishes	127,470	2	12
13 Ceiling finishes	63,735	1	6
14 Fittings and furnishings	63,735	1	6
15 Sanitary appliances/disposal installation	127,470	2	12
16 Hot and cold water services	127,470	2	12
17 Heating and air treatment installation	254,940	4	24
18 Ventilation installation	254,940	4	24
19 Gas services	127,470	2	12
20 Electric installation	637,350	10	61
21 Lift and conveyor installation	0	0	0
22 Protective & communication installation	127,470	2	12
23 Special installations/services equipment	127,470	2	12
24 Builders' work	446,145	7	42
	6,373,500	100	607

Category: Industrial
Type: Livestock building
Floor area: 800m2
Total cost: £305,600

	Elemental cost £	% of cost	Cost per m2 £
1 Preliminaries	33,616	11	42
2 Substructure	21,392	7	27
3 Frame	21,392	7	27
4 Upper floors	0	0	0
5 Roof	24,448	8	31
6 Staircases	0	0	0
7 External walls	48,896	16	61
8 Windows and external doors	9,168	3	11
9 Partitions and internal walls	15,280	5	19
10 Internal doors	3,056	1	4
11 Wall finishes	0	0	0
12 Floor finishes	9,168	3	11
13 Ceiling finishes	0	0	0
14 Fittings and furnishings	0	0	0
15 Sanitary appliances/disposal installation	0	0	0
16 Hot and cold water services	6,112	2	8
17 Heating and air treatment installation	0	0	0
18 Ventilation installation	3,056	1	4
19 Gas services	0	0	0
20 Electric installation	48,896	16	61
21 Lift and conveyor installation	0	0	0
22 Protective & communication installation	6,112	2	8
23 Special installations/services equipment	21,392	7	27
24 Builders' work	33,616	11	42
	305,600	100	382

Category: Industrial
Type: Laboratory
Floor area: 940m2
Total cost: £1,111,080

	Elemental cost £	% of cost	Cost per m2 £
1 Preliminaries	99,997	9	106
2 Substructure	66,665	6	71
3 Frame	77,776	7	83
4 Upper floors	77,776	7	83
5 Roof	88,886	8	95
6 Staircases	22,222	2	24
7 External walls	66,665	6	71
8 Windows and external doors	55,554	5	59
9 Partitions and internal walls	44,443	4	47
10 Internal doors	33,332	3	35
11 Wall finishes	22,222	2	24
12 Floor finishes	33,332	3	35
13 Ceiling finishes	33,332	3	35
14 Fittings and furnishings	44,443	4	47
15 Sanitary appliances/disposal installation	11,111	1	12
16 Hot and cold water services	11,111	1	12
17 Heating and air treatment installation	11,111	1	12
18 Ventilation installation	11,111	1	12
19 Gas services	11,111	1	12
20 Electric installation	122,219	11	130
21 Lift and conveyor installation	0	0	0
22 Protective & communication installation	22,222	2	24
23 Special installations/services equipment	33,332	3	35
24 Builders' work	111,108	10	118
	1,111,080	100	1,182

Category: Industrial
Type: Nursery units
Floor area: 2,018m2
Total cost: £1,029,180

	Elemental cost £	% of cost	Cost per m2 £
1 Preliminaries	123,502	12	61
2 Substructure	102,918	10	51
3 Frame	0	0	0
4 Upper floors	0	0	0
5 Roof	164,669	16	82
6 Staircases	0	0	0
7 External walls	226,420	22	112
8 Windows and external doors	30,875	3	15
9 Partitions and internal walls	20,584	2	10
10 Internal doors	10,292	1	5
11 Wall finishes	10,292	1	5
12 Floor finishes	20,584	2	10
13 Ceiling finishes	10,292	1	5
14 Fittings and furnishings	0	0	0
15 Sanitary appliances/disposal installation	10,292	1	5
16 Hot and cold water services	10,292	1	5
17 Heating and air treatment installation	20,584	2	10
18 Ventilation installation	0	0	0
19 Gas services	10,292	1	5
20 Electric installation	113,210	11	56
21 Lift and conveyor installation	0	0	0
22 Protective & communication installation	20,584	2	10
23 Special installations/services equipment	0	0	0
24 Builders' work	123,502	12	61
	1,029,180	100	510

Category: Industrial
Type: Warehouse shell
Floor area: 2,864m2
Total cost: £1,211,472

	Elemental cost £	% of cost	Cost per m2 £
1 Preliminaries	121,147	10	42
2 Substructure	169,606	14	59
3 Frame	193,836	16	68
4 Upper floors	12,115	1	4
5 Roof	157,491	13	55
6 Staircases	12,115	1	4
7 External walls	157,491	13	55
8 Windows and external doors	60,574	5	21
9 Partitions and internal walls	24,229	2	8
10 Internal doors	24,229	2	8
11 Wall finishes	12,115	1	4
12 Floor finishes	36,344	3	13
13 Ceiling finishes	12,115	1	4
14 Fittings and furnishings	12,115	1	4
15 Sanitary appliances/disposal installation	12,115	1	4
16 Hot and cold water services	12,115	1	4
17 Heating and air treatment installation	24,229	2	8
18 Ventilation installation	0	0	0
19 Gas services	0	0	0
20 Electric installation	84,803	7	30
21 Lift and conveyor installation	0	0	0
22 Protective & communication installation	12,115	1	4
23 Special installations/services equipment	12,115	1	4
24 Builders' work	48,459	4	17
	1,211,472	100	423

Category: Industrial
Type: Warehouse, complete
Floor area: 2,120m2
Total cost: £1,250,800

	Elemental cost £	% of cost	Cost per m2 £
1 Preliminaries	112,572	9	53
2 Substructure	137,588	11	65
3 Frame	162,604	13	77
4 Upper floors	12,508	1	6
5 Roof	162,604	13	77
6 Staircases	12,508	1	6
7 External walls	162,604	13	77
8 Windows and external doors	50,032	4	24
9 Partitions and internal walls	25,016	2	12
10 Internal doors	25,016	2	12
11 Wall finishes	25,016	2	12
12 Floor finishes	37,524	3	18
13 Ceiling finishes	25,016	2	12
14 Fittings and furnishings	25,016	2	12
15 Sanitary appliances/disposal installation	12,508	1	6
16 Hot and cold water services	12,508	1	6
17 Heating and air treatment installation	37,524	3	18
18 Ventilation installation	0	0	0
19 Gas services	12,508	1	6
20 Electric installation	87,556	7	41
21 Lift and conveyor installation	0	0	0
22 Protective & communication installation	37,524	3	18
23 Special installations/services equipment	12,508	1	6
24 Builders' work	62,540	5	30
	1,250,800	100	590

Category: Health and welfare
Type: Group surgery
Floor area: 395m2
Total cost: £411,590

	Elemental cost £	% of cost	Cost per m2 £
1 Preliminaries	49,391	12	125
2 Substructure	41,159	10	104
3 Frame	0	0	0
4 Upper floors	0	0	0
5 Roof	41,159	10	104
6 Staircases	0	0	0
7 External walls	49,391	12	125
8 Windows and external doors	24,695	6	63
9 Partitions and internal walls	24,695	6	63
10 Internal doors	12,348	3	31
11 Wall finishes	8,232	2	21
12 Floor finishes	12,348	3	31
13 Ceiling finishes	8,232	2	21
14 Fittings and furnishings	8,232	2	21
15 Sanitary appliances/disposal installation	8,232	2	21
16 Hot and cold water services	4,116	1	10
17 Heating and air treatment installation	4,116	1	10
18 Ventilation installation	4,116	1	10
19 Gas services	0	0	0
20 Electric installation	45,275	11	115
21 Lift and conveyor installation	0	0	0
22 Protective & communication installation	8,232	2	21
23 Special installations/services equipment	8,232	2	21
24 Builders' work	49,391	12	125
	411,590	100	1,042

Category: Health and welfare
Type: Health centre
Floor area: 510m2
Total cost: £626,280

	Elemental cost £	% of cost	Cost per m2 £
1 Preliminaries	68,891	11	135
2 Substructure	62,628	10	123
3 Frame	0	0	0
4 Upper floors	0	0	0
5 Roof	68,891	11	135
6 Staircases	0	0	0
7 External walls	81,416	13	160
8 Windows and external doors	25,051	4	49
9 Partitions and internal walls	25,051	4	49
10 Internal doors	18,788	3	37
11 Wall finishes	12,526	2	25
12 Floor finishes	18,788	3	37
13 Ceiling finishes	12,526	2	25
14 Fittings and furnishings	25,051	4	49
15 Sanitary appliances/disposal installation	18,788	3	37
16 Hot and cold water services	6,263	1	12
17 Heating and air treatment installation	6,263	1	12
18 Ventilation installation	6,263	1	12
19 Gas services	0	0	0
20 Electric installation	43,840	12	86
21 Lift and conveyor installation	0	0	0
22 Protective & communication installation	18,788	2	37
23 Special installations/services equipment	6,263	2	12
24 Builders' work	31,314	11	61
	626,280	100	1,228

Category: Health and welfare
Type: Old persons' nursing home
Floor area: 1,672m2
Total cost: £2,033,152

	Elemental cost £	% of cost	Cost per m2 £
1 Preliminaries	223,647	11	134
2 Substructure	162,652	8	97
3 Frame	0	0	0
4 Upper floors	0	0	0
5 Roof	264,310	13	158
6 Staircases	0	0	0
7 External walls	121,989	6	73
8 Windows and external doors	142,321	7	85
9 Partitions and internal walls	81,326	4	49
10 Internal doors	101,658	5	61
11 Wall finishes	81,326	4	49
12 Floor finishes	81,326	4	49
13 Ceiling finishes	40,663	2	24
14 Fittings and furnishings	121,989	6	73
15 Sanitary appliances/disposal installation	60,995	3	36
16 Hot and cold water services	60,995	3	36
17 Heating and air treatment installation	142,321	7	85
18 Ventilation installation	20,332	1	12
19 Gas services	20,332	1	12
20 Electric installation	121,989	6	73
21 Lift and conveyor installation	0	0	0
22 Protective & communication installation	60,995	3	36
23 Special installations/services equipment	60,995	3	36
24 Builders' work	60,995	3	36
	2,033,152	100	1,216

Category: Health and welfare
Type: Welfare centre
Floor area: 612m2
Total cost: £751,536

	Elemental cost £	% of cost	Cost per m2 £
1 Preliminaries	90,184	12	147
2 Substructure	97,700	13	160
3 Frame	0	0	0
4 Upper floors	0	0	0
5 Roof	82,669	11	135
6 Staircases	0	0	0
7 External walls	90,184	12	147
8 Windows and external doors	30,061	4	49
9 Partitions and internal walls	22,546	3	37
10 Internal doors	22,546	3	37
11 Wall finishes	22,546	3	37
12 Floor finishes	22,546	3	37
13 Ceiling finishes	15,031	2	25
14 Fittings and furnishings	37,577	5	61
15 Sanitary appliances/disposal installation	15,031	2	25
16 Hot and cold water services	7,515	1	12
17 Heating and air treatment installation	15,031	2	25
18 Ventilation installation	7,515	1	12
19 Gas services	0	0	0
20 Electric installation	82,669	11	135
21 Lift and conveyor installation	0	0	0
22 Protective & communication installation	15,031	2	25
23 Special installations/services equipment	7,515	1	12
24 Builders' work	67,638	9	111
	751,536	100	1,228

Category: Leisure
Type: Golf club house
Floor area: 675m2
Total cost: £553,500

	Elemental cost £	% of cost	Cost per m2 £
1 Preliminaries	55,350	10	82
2 Substructure	38,745	7	57
3 Frame	0	0	0
4 Upper floors	0	0	0
5 Roof	77,490	14	115
6 Staircases	0	0	0
7 External walls	60,885	11	90
8 Windows and external doors	22,140	4	33
9 Partitions and internal walls	16,605	3	25
10 Internal doors	16,605	3	25
11 Wall finishes	16,605	3	25
12 Floor finishes	11,070	2	16
13 Ceiling finishes	11,070	2	16
14 Fittings and furnishings	44,280	8	66
15 Sanitary appliances/disposal installation	11,070	2	16
16 Hot and cold water services	5,535	1	8
17 Heating and air treatment installation	5,535	1	8
18 Ventilation installation	0	0	0
19 Gas services	5,535	1	8
20 Electric installation	77,490	14	115
21 Lift and conveyor installation	0	0	0
22 Protective & communication installation	11,070	2	16
23 Special installations/services equipment	5,535	1	8
24 Builders' work	60,885	11	90
	553,500	100	820

Category: Leisure
Type: Public house
Floor area: 780m2
Total cost: £791,700

	Elemental cost £	% of cost	Cost per m2 £
1 Preliminaries	71,253	9	91
2 Substructure	47,502	6	61
3 Frame	15,834	2	20
4 Upper floors	15,834	2	20
5 Roof	79,170	10	102
6 Staircases	23,751	3	30
7 External walls	47,502	6	61
8 Windows and external doors	31,668	4	41
9 Partitions and internal walls	23,751	3	30
10 Internal doors	31,668	4	41
11 Wall finishes	39,585	5	51
12 Floor finishes	39,585	5	51
13 Ceiling finishes	31,668	4	41
14 Fittings and furnishings	95,004	12	122
15 Sanitary appliances/disposal installation	23,751	3	30
16 Hot and cold water services	7,917	1	10
17 Heating and air treatment installation	31,668	4	41
18 Ventilation installation	7,917	1	10
19 Gas services	7,917	1	10
20 Electric installation	55,419	7	71
21 Lift and conveyor installation	0	0	0
22 Protective & communication installation	15,834	2	20
23 Special installations/services equipment	31,668	4	41
24 Builders' work	15,834	2	20
	791,700	100	1,015

Category: Leisure
Type: Restaurant
Floor area: 340m2
Total cost: £404,600

	Elemental cost £	% of cost	Cost per m2 £
1 Preliminaries	36,414	9	107
2 Substructure	32,368	8	95
3 Frame	0	0	0
4 Upper floors	0	0	0
5 Roof	44,506	11	131
6 Staircases	0	0	0
7 External walls	52,598	13	155
8 Windows and external doors	24,276	6	71
9 Partitions and internal walls	28,322	7	83
10 Internal doors	24,276	6	71
11 Wall finishes	16,184	4	48
12 Floor finishes	20,230	5	60
13 Ceiling finishes	12,138	3	36
14 Fittings and furnishings	28,322	7	83
15 Sanitary appliances/disposal installation	4,046	1	12
16 Hot and cold water services	4,046	1	12
17 Heating and air treatment installation	4,046	1	12
18 Ventilation installation	4,046	1	12
19 Gas services	4,046	1	12
20 Electric installation	56,644	10	167
21 Lift and conveyor installation	0	0	0
22 Protective & communication installation	8,092	2	24
23 Special installations/services equipment	4,046	1	12
24 Builders' work	44,506	3	131
	404,600	100	1,190

Category: Leisure
Type: Sports hall
Floor area: 2,016m2
Total cost: £1,465,632

	Elemental cost £	% of cost	Cost per m2 £
1 Preliminaries	87,938	6	44
2 Substructure	146,563	10	73
3 Frame	175,876	12	87
4 Upper floors	0	0	0
5 Roof	161,220	11	80
6 Staircases	0	0	0
7 External walls	102,594	7	51
8 Windows and external doors	43,969	3	22
9 Partitions and internal walls	43,969	3	22
10 Internal doors	29,313	2	15
11 Wall finishes	29,313	2	15
12 Floor finishes	102,594	7	51
13 Ceiling finishes	14,656	1	7
14 Fittings and furnishings	117,251	8	58
15 Sanitary appliances/disposal installation	58,625	4	29
16 Hot and cold water services	29,313	2	15
17 Heating and air treatment installation	58,625	4	29
18 Ventilation installation	14,656	1	7
19 Gas services	0	0	0
20 Electric installation	117,251	8	58
21 Lift and conveyor installation	0	0	0
22 Protective & communication installation	29,313	2	15
23 Special installations/services equipment	14,656	1	7
24 Builders' work	87,938	6	44
	1,465,632	100	727

Category: Education
Type: Library
Floor area: 616m2
Total cost: £607,992

	Elemental cost £	% of cost	Cost per m2 £
1 Preliminaries	72,959	12	118
2 Substructure	54,719	9	89
3 Frame	0	0	0
4 Upper floors	0	0	0
5 Roof	72,959	12	118
6 Staircases	0	0	0
7 External walls	85,119	14	138
8 Windows and external doors	42,559	7	69
9 Partitions and internal walls	12,160	2	20
10 Internal doors	12,160	2	20
11 Wall finishes	18,240	3	30
12 Floor finishes	18,240	3	30
13 Ceiling finishes	12,160	2	20
14 Fittings and furnishings	54,719	9	89
15 Sanitary appliances/disposal installation	6,080	1	10
16 Hot and cold water services	6,080	1	10
17 Heating and air treatment installation	12,160	2	20
18 Ventilation installation	0	0	0
19 Gas services	0	0	0
20 Electric installation	48,639	8	79
21 Lift and conveyor installation	0	0	0
22 Protective & communication installation	12,160	2	20
23 Special installations/services equipment	6,080	1	10
24 Builders' work	60,799	10	99
	607,992	100	987

Category: Education
Type: Primary school
Floor area: 2,108m2
Total cost: £2,238,696

	Elemental cost £	% of cost	Cost per m2 £
1 Preliminaries	201,483	9	96
2 Substructure	156,709	7	74
3 Frame	0	0	0
4 Upper floors	0	0	0
5 Roof	223,870	10	106
6 Staircases	0	0	0
7 External walls	246,257	11	117
8 Windows and external doors	268,644	12	127
9 Partitions and internal walls	134,322	6	64
10 Internal doors	89,548	4	42
11 Wall finishes	67,161	3	32
12 Floor finishes	89,548	4	42
13 Ceiling finishes	44,774	2	21
14 Fittings and furnishings	134,322	6	64
15 Sanitary appliances/disposal installation	89,548	4	42
16 Hot and cold water services	22,387	1	11
17 Heating and air treatment installation	89,548	4	42
18 Ventilation installation	0	0	0
19 Gas services	22,387	1	11
20 Electric installation	179,096	8	85
21 Lift and conveyor installation	0	0	0
22 Protective & communication installation	44,774	2	21
23 Special installations/services equipment	22,387	1	11
24 Builders' work	111,935	5	53
	2,238,696	100	1,062

Category: Education
Type: Secondary school
Floor area: 2,450m2
Total cost: £2,410,800

	Elemental cost £	% of cost	Cost per m2 £
1 Preliminaries	216,972	9	89
2 Substructure	168,756	7	69
3 Frame	0	0	0
4 Upper floors	0	0	0
5 Roof	265,188	11	108
6 Staircases	0	0	0
7 External walls	241,080	10	98
8 Windows and external doors	216,972	9	89
9 Partitions and internal walls	120,540	5	49
10 Internal doors	96,432	4	39
11 Wall finishes	72,324	3	30
12 Floor finishes	96,432	4	39
13 Ceiling finishes	48,216	2	20
14 Fittings and furnishings	216,972	9	89
15 Sanitary appliances/disposal installation	48,216	2	20
16 Hot and cold water services	24,108	1	10
17 Heating and air treatment installation	72,324	3	30
18 Ventilation installation	0	0	0
19 Gas services	0	0	0
20 Electric installation	216,972	9	89
21 Lift and conveyor installation	0	0	0
22 Protective & communication installation	48,216	2	20
23 Special installations/services equipment	24,108	1	10
24 Builders' work	216,972	9	89
	2,410,800	100	984

Category: Education
Type: Sixth form college
Floor area: 1,828m2
Total cost: £2,047,360

	Elemental cost £	% of cost	Cost per m2 £
1 Preliminaries	204,736	10	112
2 Substructure	163,789	8	90
3 Frame	0	0	0
4 Upper floors	0	0	0
5 Roof	184,262	9	101
6 Staircases	0	0	0
7 External walls	245,683	12	134
8 Windows and external doors	163,789	8	90
9 Partitions and internal walls	102,368	5	56
10 Internal doors	81,894	4	45
11 Wall finishes	61,421	3	34
12 Floor finishes	61,421	3	34
13 Ceiling finishes	40,947	2	22
14 Fittings and furnishings	245,683	12	134
15 Sanitary appliances/disposal installation	20,474	1	11
16 Hot and cold water services	20,474	1	11
17 Heating and air treatment installation	61,421	3	34
18 Ventilation installation	0	0	0
19 Gas services	0	0	0
20 Electric installation	163,789	8	90
21 Lift and conveyor installation	0	0	0
22 Protective & communication installation	40,947	2	22
23 Special installations/services equipment	20,474	1	11
24 Builders' work	163,789	8	90
	2,047,360	100	1,120

Category: Education
Type: Special school
Floor area: 1,710m2
Total cost: £1,839,960

	Elemental cost £	% of cost	Cost per m2 £
1 Preliminaries	202,396	11	118
2 Substructure	165,596	9	97
3 Frame	0	0	0
4 Upper floors	0	0	0
5 Roof	183,996	10	108
6 Staircases	0	0	0
7 External walls	202,396	11	118
8 Windows and external doors	128,797	7	75
9 Partitions and internal walls	91,998	5	54
10 Internal doors	55,199	3	32
11 Wall finishes	55,199	3	32
12 Floor finishes	55,199	3	32
13 Ceiling finishes	36,799	2	22
14 Fittings and furnishings	257,594	14	151
15 Sanitary appliances/disposal installation	36,799	2	22
16 Hot and cold water services	18,400	1	11
17 Heating and air treatment installation	55,199	3	32
18 Ventilation installation	0	0	0
19 Gas services	0	0	0
20 Electric installation	147,197	8	86
21 Lift and conveyor installation	0	0	0
22 Protective & communication installation	36,799	2	22
23 Special installations/services equipment	36,799	2	22
24 Builders' work	73,598	4	43
	1,839,960	100	1,076

Category: Education
Type: Teachers' training college
Floor area: 2,620m2
Total cost: £2,520,440

	Elemental cost £	% of cost	Cost per m2 £
1 Preliminaries	252,044	10	96
2 Substructure	201,635	8	77
3 Frame	151,226	6	58
4 Upper floors	100,818	4	38
5 Roof	252,044	10	96
6 Staircases	75,613	3	29
7 External walls	226,840	9	87
8 Windows and external doors	151,226	6	58
9 Partitions and internal walls	100,818	4	38
10 Internal doors	100,818	4	38
11 Wall finishes	75,613	3	29
12 Floor finishes	75,613	3	29
13 Ceiling finishes	50,409	2	19
14 Fittings and furnishings	176,431	7	67
15 Sanitary appliances/disposal installation	50,409	2	19
16 Hot and cold water services	25,204	1	10
17 Heating and air treatment installation	50,409	2	19
18 Ventilation installation	0	0	0
19 Gas services	0	0	0
20 Electric installation	201,635	8	77
21 Lift and conveyor installation	0	0	0
22 Protective & communication installation	50,409	2	19
23 Special installations/services equipment	25,204	1	10
24 Builders' work	126,022	5	48
	2,520,440	100	962

Category: Residential
Type: Local Authority low rise flats
Floor area: 2,349m2
Total cost: £1,785,240

	Elemental cost £	% of cost	Cost per m2 £
1 Preliminaries	178,524	10	76
2 Substructure	178,524	10	76
3 Frame	0	0	0
4 Upper floors	71,410	4	30
5 Roof	196,376	11	84
6 Staircases	35,705	2	15
7 External walls	214,229	12	91
8 Windows and external doors	142,819	8	61
9 Partitions and internal walls	89,262	5	38
10 Internal doors	71,410	4	30
11 Wall finishes	53,557	3	23
12 Floor finishes	53,557	3	23
13 Ceiling finishes	89,262	5	38
14 Fittings and furnishings	53,557	3	23
15 Sanitary appliances/disposal installation	35,705	2	15
16 Hot and cold water services	17,852	1	8
17 Heating and air treatment installation	107,114	6	46
18 Ventilation installation	17,852	1	8
19 Gas services	0	0	0
20 Electric installation	89,262	5	38
21 Lift and conveyor installation	0	0	0
22 Protective & communication installation	35,705	2	15
23 Special installations/services equipment	35,705	2	15
24 Builders' work	17,852	1	8
	1,785,240	100	760

Category: Residential
Type: Local Authority housing
Floor area: 3,988m2
Total cost: £2,759,696

	Elemental cost £	% of cost	Cost per m2 £
1 Preliminaries	248,373	9	62
2 Substructure	165,582	6	42
3 Frame	0	0	0
4 Upper floors	193,179	7	48
5 Roof	358,760	13	90
6 Staircases	82,791	3	21
7 External walls	275,970	10	69
8 Windows and external doors	165,582	6	42
9 Partitions and internal walls	110,388	4	28
10 Internal doors	82,791	3	21
11 Wall finishes	110,388	4	28
12 Floor finishes	55,194	2	14
13 Ceiling finishes	55,194	2	14
14 Fittings and furnishings	110,388	4	28
15 Sanitary appliances/disposal installation	55,194	2	14
16 Hot and cold water services	55,194	2	14
17 Heating and air treatment installation	82,791	3	21
18 Ventilation installation	0	0	0
19 Gas services	0	0	0
20 Electric installation	386,357	14	97
21 Lift and conveyor installation	0	0	0
22 Protective & communication installation	0	0	0
23 Special installations/services equipment	0	0	0
24 Builders' work	165,582	6	42
	2,759,696	100	692

Category: Residential
Type: Private flats
Floor area: 1,260m2
Total cost: £1,020,600

	Elemental cost £	% of cost	Cost per m2 £
1 Preliminaries	61,236	6	49
2 Substructure	61,236	6	49
3 Frame	0	0	0
4 Upper floors	51,030	5	41
5 Roof	122,472	12	97
6 Staircases	30,618	3	24
7 External walls	112,266	11	89
8 Windows and external doors	61,236	6	49
9 Partitions and internal walls	30,618	3	24
10 Internal doors	40,824	4	32
11 Wall finishes	30,618	3	24
12 Floor finishes	30,618	3	24
13 Ceiling finishes	30,618	3	24
14 Fittings and furnishings	40,824	4	32
15 Sanitary appliances/disposal installation	10,206	1	8
16 Hot and cold water services	20,412	2	16
17 Heating and air treatment installation	30,618	3	24
18 Ventilation installation	0	0	0
19 Gas services	0	0	0
20 Electric installation	112,266	11	89
21 Lift and conveyor installation	40,824	4	32
22 Protective & communication installation	30,618	3	24
23 Special installations/services equipment	10,206	1	8
24 Builders' work	61,236	6	49
	1,020,600	100	810

Category: Residential
Type: Luxury private flats
Floor area: 2,400m2
Total cost: £2,832,000

	Elemental cost £	% of cost	Cost per m2 £
1 Preliminaries	198,240	7	83
2 Substructure	169,920	6	71
3 Frame	0	0	0
4 Upper floors	198,240	7	83
5 Roof	254,880	9	106
6 Staircases	84,960	3	35
7 External walls	283,200	10	118
8 Windows and external doors	198,240	7	83
9 Partitions and internal walls	84,960	3	35
10 Internal doors	113,280	4	47
11 Wall finishes	84,960	3	35
12 Floor finishes	56,640	2	24
13 Ceiling finishes	56,640	2	24
14 Fittings and furnishings	169,920	6	71
15 Sanitary appliances/disposal installation	56,640	2	24
16 Hot and cold water services	56,640	2	24
17 Heating and air treatment installation	84,960	3	35
18 Ventilation installation	0	0	0
19 Gas services	0	0	0
20 Electric installation	283,200	10	118
21 Lift and conveyor installation	141,600	5	59
22 Protective & communication installation	84,960	3	35
23 Special installations/services equipment	28,320	1	12
24 Builders' work	141,600	5	59
	2,832,000	100	1,180

Category: Transport
Type: Multi-storey car park
Floor area: 9,600m2
Total cost: £2,976,000

	Elemental cost £	% of cost	Cost per m2 £
1 Preliminaries	327,360	11	34
2 Substructure	386,880	13	40
3 Frame	624,960	21	65
4 Upper floors	535,680	18	56
5 Roof	29,760	1	3
6 Staircases	148,800	5	16
7 External walls	119,040	4	12
8 Windows and external doors	29,760	1	3
9 Partitions and internal walls	29,760	1	3
10 Internal doors	29,760	1	3
11 Wall finishes	0	0	0
12 Floor finishes	0	0	0
13 Ceiling finishes	0	0	0
14 Fittings and furnishings	0	0	0
15 Sanitary appliances/disposal installation	0	0	0
16 Hot and cold water services	0	0	0
17 Heating and air treatment installation	0	0	0
18 Ventilation installation	0	2	0
19 Gas services	0	0	0
20 Electric installation	267,840	9	28
21 Lift and conveyor installation	208,320	7	22
22 Protective & communication installation	89,280	3	9
23 Special installations/services equipment	59,520	2	6
24 Builders' work	29,760	1	3
	2,976,000	100	304

MECHANICAL AND ELECTRICAL ELEMENTAL COSTS

	Office £/m2	%	Hospital £/m2	%	Hotel £/m2	%
Mechanical work						
Heating	80	50	105	48	91	48
Ventilation	49	30	71	33	53	28
Fire protection	23	14	26	12	35	28
Ancillary services	9	6	13	7	11	6
	161	100	215	100	190	100

	Office £/m2	%	Hospital £/m2	%	Hotel £/m2	%
Electrical work						
Lighting	43	37	67	35	83	35
Mains	29	24	53	28	81	34
Power	23	19	26	14	35	15
Lightning protection	5	4	5	3	5	2
Communications	5	4	16	8	14	6
Security	6	5	14	8	13	5
Ancillary services	8	7	8	4	8	3
	119	100	189	100	239	100

	Leisure		Stadium		Industrial	
	£/m2	%	£/m2	%	£/m2	%
Mechanical work						
Heating	36	25	3	15	38	45
Ventilation	73	52	3	15	20	24
Fire protection	22	16	10	50	14	17
Ancillary services	10	7	4	20	12	14
	141	100	20	100	84	100

	Leisure		Stadium		Industrial	
	£/m2	%	£/m2	%	£/m2	%
Electrical work						
Lighting	44	36	31	33	28	27
Mains	27	22	17	19	22	22
Power	18	15	11	12	21	21
Lightning protection	3	2	4	4	3	3
Communications	8	6	10	10	5	5
Security	12	10	11	12	12	12
Ancillary services	10	9	10	10	9	9
	122	100	94	100	100	100

4

Composite and principal rates

When the appraisal of a project moves beyond the square metre stage, it is likely that the client will commission a cost plan to be prepared. This entails taking off rough quantities and applying global rates to produce an approximate cost of the project. These rates are the result of combining various item descriptions and costs into what are called composite rates.

Not all items can be combined with others to provide composite rates so principal rates are also included if their value is significant to the cost plan. The following rates are presented under headings in the same order as the elements in Chapter 3.

BUILDING WORK

PRELIMINARIES

These will usually be assessed by referring to the needs of each particular project but a figure of 7.5 to 12.5% of the construction costs is normal depending upon the nature of the project.

SUBSTRUCTURES

Excavation		£
Excavate by machine to reduce levels and dispose of excavated material		
deposit on site	m3	6.00
spread and level on site	m3	7.00
remove from site	m3	20.00
Excavate by machine for basements and dispose of excavated material		
deposit on site	m3	7.00
spread and level on site	m3	8.00
remove from site	m3	20.00

£

Breaking out and disposal

Extra over excavation for breaking out

rock	m3	45.00
concrete	m3	35.00
reinforced concrete	m3	55.00
brickwork or blockwork	m3	30.00

Extra over excavation for disposing of contaminated material to licensed tip	m3	50.00

Filling

Filling to make up levels including
levelling and compacting

excavated material	m3	6.00
sand	m3	35.00
hardcore	m3	30.00
DOT Type 1	m3	35.00
DOT Type 2	m3	33.00

Foundation walling

Excavate trench by machine, dispose of
surplus excavated material off site,
earthwork support, concrete foundations
10 N/mm2 – 40mm aggregate (1:3:6),
common bricks (£150/1000) in cavity wall,
pitch polymer damp proof course,
concrete foundation size

450 x 150mm

wall height 750mm	m	110.00
wall height 1000mm	m	120.00
wall height 1250mm	m	130.00

£

450 x 225mm

wall height 750mm	m	116.00
wall height 1000mm	m	126.00
wall height 1250mm	m	136.00

600 x 225mm

wall height 750mm	m	122.00
wall height 1000mm	m	132.00
wall height 1250mm	m	145.00

Excavate trench by machine, dispose of
surplus excavated material off site,
earthwork support, concrete foundations
10 N/mm2 – 40mm aggregate (1:3:6),
engineering bricks (£200/1000) in cavity wall
and damp proof course for foundation size

450 x 150mm

wall height 750mm	m	116.00
wall height 1000mm	m	126.00
wall height 1250mm	m	136.00

450 x 225mm

wall height 750mm	m	122.00
wall height 1000mm	m	126.00
wall height 1250mm	m	142.00

600 x 225mm

wall height 750mm	m	128.00
wall height 1000mm	m	138.00
wall height 1250mm	m	150.00

£

Bases, pile caps and ground beams

Reinforced concrete (11.5N/mm2 40mm
aggregate) in bases including excavation,
earthwork support, formwork and
reinforcement, size

600 x 600 x 300mm	nr	40.00
750 x 750 x 450	nr	56.00
1000 x 1000 x 450mm	nr	90.00
1200 x 1200 x 600mm	nr	172.00

Reinforced concrete (21N/mm2 20mm
aggregate) in pile caps including
excavation, earthwork support,
formwork and reinforcement, size

750 x 750 x 600mm	nr	70.00
900 x 900 x 1000mm	nr	130.00
1000 x 1000 x 1000mm	nr	200.00
2000 x 2000 x 1000mm	nr	800.00

Reinforced concrete (21N/mm2 20mm
aggregate) in ground beams
including excavation, earthwork
support, formwork and reinforcement,
size

450 x 450	m	40.00
750 x 500	m	75.00
900 x 500	m	90.00
1000 x 1000	m	200.00

£

Strip foundations and slabs

Strip foundations including trench
excavation, disposal, earthwork
support, concrete foundation (11.5N/m2
40mm aggregate), brick cavity wall with
three courses of facing bricks, wall height

in commons (£150/1000)			
600mm	m	90.00
900mm	m	116.00
1200mm	m	140.00
1500mm	m	180.00
in facings (£250/1000)			
600mm	m	100.00
900mm	m	130.00
1200mm	m	160.00
1500mm	m	210.00
in engineering bricks (230/1000)			
600mm	m	96.00
900mm	m	120.00
1200mm	m	150.00
1500mm	m	200.00

Ground slab including excavation,
disposal, hardcore bed blinded with
with sand, 1200 gauge polythene damp proof
membrane, concrete (21Nmm2 20mm
aggregate)

one layer of reinforcement A252			
150mm thick	m2	42.00
200mm "	m2	44.00
250mm "	m2	46.00
300mm "	m2	48.00

		£

two layers of reinforcement A252

150mm	m2	44.00
200mm	m2	46.00
250mm	m2	48.00
300mm	m2	50.00

FRAME

Reinforced concrete (30N/mm2
20mm aggregate) including
reinforcement and formwork

columns, size

225 x 225mm	m	48.00
300 x 300mm	m	66.00
300 x 450mm	m	86.00
450 x 450mm	m	124.00
450 x 600mm	m	150.00
450 x 900mm	m	200.00

Reinforced concrete (30N/mm2
20mm aggregate) including
reinforcement and formwork

beams, size

225 x 225mm	m	58.00
300 x 450mm	m	100.00
300 x 600mm	m	124.00
450 x 450mm	m	134.00
450 x 600mm	m	148.00
450 x 900mm	m	190.00
600 x 900mm	m	230.00
600 x 1200mm	m	300.00

£

Fabricated galvanised steelwork
BS4360 grade 40 erected on site
Including connections

columns	t	2,250.00
universal beams	t	2,750.00
roof bracing	t	2,250.00
lattice beams	t	2,500.00

UPPER FLOORS

Softwood flooring including 25mm
thick tongued and grooved boarding
on joists size

125 x 50mm	m2	32.00
150 x 50mm	m2	34.00
175 x 50mm	m2	36.00
200 x 50mm	m2	38.00
225 x 50mm	m2	40.00
250 x 50mm	m2	42.00

Softwood flooring including 18mm
thick chipboard flooring butt jointed
on joists size

125 x 50mm	m2	28.00
150 x 50mm	m2	30.00
175 x 50mm	m2	32.00
200 x 50mm	m2	34.00
225 x 50mm	m2	36.00
250 x 50mm	m2	38.00

Reinforced concrete suspended slabs
Including formwork

150mm thick	m2	80.00
200mm thick	m2	100.00

		£
Precast concrete suspended slabs		
150mm thick	m2	95.00
200mm thick	m2	110.00

ROOFS

The following items represent
the costs of roofs measured on
plan. The costs for roof coverings
are given separately.

Flat roofs

Reinforced concrete suspended slabs including formwork		
150mm thick	m2	75.00
200mm thick	m2	90.00

Precast concrete suspended slabs		
150mm thick	m2	100.00
200mm thick	m2	110.00

Softwood flat roofing consisting of
herringbone strutting, wall plates,
joists and woodwool slabs

woodwool slabs 50mm thick		
joists 150 x 50mm	m2	40.00
joists 175 x 50mm	m2	42.00
joists 200 x 50mm	m2	44.00
joists 250 x 50mm	m2	46.00

woodwool slabs 75mm thick		
joists 150 x 50mm	m2	46.00
joists 175 x 50mm	m2	48.00
joists 200 x 50mm	m2	50.00
joists 250 x 50mm	m2	52.00

£

woodwool slabs 100mm thick

joists 150 x 50mm	m2	56.00
joists 175 x 50mm	m2	58.00
joists 200 x 50mm	m2	60.00
joists 250 x 50mm	m2	62.00

Flat roof decking

Reinforced woodwool slab decking
50mm thick covered with

two layer bituminous roofing	m2	46.00
three layer bituminous roofing	m2	50.00
two layer bituminous roofing	m2	54.00

Reinforced woodwool slab decking
75mm thick covered with

two layer bituminous roofing	m2	58.00
three layer bituminous roofing	m2	62.00
two layer bituminous roofing	m2	66.00

Reinforced woodwool slab decking
100mm thick covered with

two layer bituminous roofing	m2	70.00
three layer bituminous roofing	m2	74.00
two layer bituminous roofing	m2	78.00

Pitched roofs

Softwood trusses at 600mm centres

span 5m

pitch 22.5 degrees	m2	28.00
pitch 30 degrees	m2	32.00
pitch 40 degrees	m2	38.00
pitch 45 degrees	m2	42.00

£

span 8m

pitch 22.5 degrees	m2	32.00
pitch 30 degrees	m2	36.00
pitch 40 degrees	m2	40.00
pitch 45 degrees	m2	44.00

span 10m

pitch 22.5 degrees	m2	36.00
pitch 30 degrees	m2	40.00
pitch 40 degrees	m2	44.00
pitch 45 degrees	m2	48.00

Underlays

Roofing felt

unreinforced	m2	2.00
reinforced	m2	3.00

Building paper	m2	2.50

External quality plywood

12mm thick	m2	20.00
15mm thick	m2	22.00
18mm thick	m2	24.00

Chipboard

12mm thick	m2	7.00
15mm thick	m2	8.00
18mm thick	m2	10.00
22mm thick	m2	12.00

Insulation quilt

100mm thick	m2	8.00
150mm thick	m2	12.00
200mm thick	m2	16.00

£

Roof coverings

The following costs include for underfelt, roof battens and work to eaves, verges and ridges. The rates are based on sloping areas.

Welsh blue slates size

610 x 305mm	m2	100.00
510 x 255mm	m2	95.00
405 x 205mm	m2	96.00

Westmoreland green slates size

610 x 305mm	m2	160.00
405 x 255mm	m2	150.00
355 x 205mm	m2	155.00

Reconstructed stone slates size

457 x 305mm	m2	55.00
457 x 457mm	m2	50.00

Asbestos-free slates

400 x 200mm	m2	65.00
500 x 250mm	m2	60.00
600 x 300mm	m2	55.00

Concrete interlocking tiles

400 x 330mm	m2	28.00
380 x 230mm	m2	32.00
325 x 330mm	m2	30.00
420 x 220mm	m2	30.00
430 x 380mm	m2	24.00

£

Clay pantiles

340 x 240mm	m2	42.00.
470 x 280mm	m2	38.00.

Eaves, fascias and verges

Painted softwood fascia 250mm
high and 16mm thick softwood
soffit

200mm wide	m	30.00
400mm wide	m	32.00

Painted softwood fascia 250mm
high and 12mm thick plywood
soffit

200mm wide	m	28.00
400mm wide	m	30.00

Painted softwood barge board
250mm high to sloping roof with
painted softwood soffit

150mm wide	m	24.00
250mm wide	m	28.00

Rainwater goods

The following costs include
for all fittings.

Rainwater pipes

Cast iron		
75mm	m	36.00
100mm	m	44.00

£

PVC-U
	68mm	m	16.00
	112mm	m	20.00

aluminium
	63mm	m	28.00
	75mm	m	32.00
	100mm	m	44.00

Rainwater gutters

Cast iron
	100mm	m	30.00
	125mm	m	34.00

PVC-U
	75mm	m	16.00
	112mm	m	20.00

aluminium
	100mm	m	28.00
	112mm	m	32.00
	125mm	m	36.00

Sheet coverings

Lead sheeting
	code 5	m2	90.00
	code 6	m2	95.00

Aluminium sheeting
	0.60mm thick	m2	65.00
	0.90 thick	m2	70.00

		£
Copper sheeting		
0.56mm thick	m2	85.00
0.61mm thick	m2	90.00
Zinc sheeting		
0.65mm thick	m2	80.00
0.80mm thick	m2	85.00
Bituminous built up roofing		
two layer	m2	16.00
three layer	m2	22.00
Two coat asphalt		
13mm thick	m2	16.00
20mm thick	m2	20.00
30mm thick	m2	26.00

Roof cladding

Asbestos-free corrugated single skin cladding		
grey	m2	48.00
coloured	m2	52.00
Lightweight galvanised steel sheeting	m2	28.0
Aluminium profiled cladding with pre-painted finish	m2	38.00

£

STAIRCASES

Reinforced concrete construction,
3250mm rise, granolithic finish
including mild steel painted
balustrade and handrails

Straight flight, width

900mm	nr	3,500
1200mm	nr	3,950

Dog-leg flight, width

900mm	nr	3,800
1200mm	nr	4,250

Reinforced concrete construction,
3250mm rise, terrazzo finish
including mild steel painted
balustrade and handrails

Straight flight, width

900mm	nr	3,750
1200mm	nr	4,300

Dog-leg flight, width

900mm	nr	4,200
1200mm	nr	4,600

Softwood construction, 2600mm rise

straight flight 900mm wide, no balustrade	nr	1,200
two flights with quarter landing softwood balustrade	nr	1,450
two flights with half landing hardwood balustrade	nr	1,750

£

Mild steel construction, 3000mm rise
(balustrades and handrails excluded)

straight flight 900mm wide	nr	1,100
two flights with quarter landing	nr	1,350
two flights with half landing	nr	1,500

Spiral staircase in steel including
powder coated mild steel
balustrades and handrail, 3500mm
overall rise

1500mm diameter	nr	2,650
2000mm diameter	nr	3,000

EXTERNAL WALLS

Solid brickwork/blockwork

Common bricks £120 per 1000

half brick thick	m2	50.00
one brick thick	m2	90.00
one and a half brick thick	m2	115.00

Common bricks £150 per 1000

half brick thick	m2	55.00
one brick thick	m2	90.00
one and a half brick thick	m2	130.00

Facing bricks £300 per 1000

half brick thick	m2	65.00
one brick thick	m2	120.00

Facing bricks £450 per 1000

half brick thick	m2	90.00

		£
Engineering bricks £300 per 1000		
half brick thick	m2	70.00
one brick thick	m2	125.00
one and a half brick thick	m2	160.00
Engineering bricks £400 per 1000		
half brick thick	m2	80.00
one brick thick	m2	150.00
one and a half brick thick	m2	180.00
Lightweight concrete blocks in wall		
100mm thick	m2	26.00
140mm thick	m2	
Medium dense blocks in walls		
190mm thick	m2	38.00
75mm thick	m2	24.00
100mm thick	m2	30.00
150mm thick	m2	36.00
Dense concrete blocks in wall		
100mm thick	m2	30.00
140mm thick	m2	40.00
Reconstructed stone blocks in wall		
100mm thick	m2	55.00
Concrete (21N/m 20mm aggregate) in wall **Including reinforcement**		
sawn formwork		
100mm thick	m2	100.00
150mm thick	m2	110.00
200mm thick	m2	120.00
250mm thick	m2	130.00
300mm thick	m2	140.00

£

Concrete walls

wrought formwork			
	100mm thick	m2	110.00
	150mm thick	m2	120.00
	200mm thick	m2	130.00
	250mm thick	m2	140.00
	300mm thick	m2	150.00

Composite walls

Cavity wall formed with
lightweight block inner skin
100mm thick, stainless steel
wall ties, 50mm thick insulation
and outer skin of

common brick 102mm thick (£120 per 1000)	m2	75.00
common brick 102mm thick (£150 per 1000)	m2	80.00
facing brickwork 102mm thick (£300 per 1000)	m2	90.00
facing brickwork 102mm thick (£450 per 1000)	m2	105.00
engineering brickwork 102mm thick (£300 per 1000)	m2	95.00
engineering brickwork 102mm thick (£400 per 1000)	m2	105.00
reconstructed stone block 100mm thick	m2	80.00
in situ concrete wall 100mm thick	m2	125.00
in situ concrete wall 150mm thick	m2	135.00
in situ concrete wall 200mm thick	m2	145.00
in situ concrete wall 250mm thick	m2	155.00
in situ concrete wall 300mm thick	m2	165.00

£

Cavity wall insulation

25mm thick board	m2	12.00
50mm thick board	m2	15.00

Wall cladding

Precast concrete panels	m2	170.00
Precast concrete panels with exposed aggregate finish	m2	220.00
Precast concrete panels with reconstituted stone facing	m2	450.00
Precast concrete panels with Portland stone natural facing	m2	575.00

GRP panels including associated
fixings, insulation and flashings

single skin panels	m2	200.00
double skin panels	m2	240.00

Non-asbestos coloured profiled cladding including insulation	m2	36.00

Metal cladding systems

PVF2 coated profiled galvanised steel sheeting, 80mm thick insulation and coated inner lining	m2	56.00
Insulated PVF2 coated silver finished metal sandwich panel system	m2	96.00

£

Curtain walling

Single glazed powder coated aluminium framed curtain walling system	m2	360.00
Double glazed powder coated aluminium framed curtain walling system	m2	480.00

External finishes

Two coat render with painted finish	m2	28.00
Two coated masonry paint	m2	14.00
Clay tile hanging including battens and felt	m2	36.00
Boarding in Western Red Cedar	m2	34.00

WINDOWS AND EXTERNAL DOORS

Windows

Softwood painted windows

single glazed	m2	220.00
double glazed	m2	250.00

Hardwood painted windows

single glazed	m2	340.00
double glazed	m2	370.00

Steel painted windows

single glazed	m2	240.00
double glazed	m2	270.00

£

Aluminium painted windows

single glazed	m2	320.00
double glazed	m2	290.00

PVC-U windows

double glazed	m2	420.00

External doors (including frames and ironmongery)

Softwood framed, ledged and braced door	nr	220.00
Softwood flush door	nr	270.00
Softwood panelled door	nr	340.00
Hardwood flush door	nr	500.00
Hardwood four panelled door	nr	600.00

Steel roller shutters and grilles

Manual	m2	300.00
Electric	m2	420.00

PARTITIONS AND INTERNAL WALLS

Common bricks £150 per 1000

half brick thick	m2	55.00
one brick thick	m2	90.00

£

Lightweight concrete blocks in wall

100mm thick	m2	24.00
140mm thick	m2	30.00
190mm thick	m2	36.00

Medium dense concrete blocks in wall

75mm thick	m2	26.00
100mm thick	m2	32.00
150mm thick	m2	38.00

Dense concrete blocks in wall

100mm thick	m2	30.00
140mm thick	m2	40.00

Concrete (21N/m 20mm aggregate) in wall
including reinforcement

sawn formwork

100mm thick	m2	100.00
150mm thick	m2	110.00

wrought formwork

100mm thick	m2	110.00
150mm thick	m2	120.00

Stud partition faced with

plasterboard 9.5mm thick

one side	m2	40.00
two sides	m2	48.00

plasterboard 12.5mm thick

one side	m2	50.00
two sides	m2	58.00

£

Extra for

skim coat plaster		
one side	m2	7.00
two sides	m2	14.00
one mist and two coats of emulsion		
one side	m2	6.00
two sides	m2	12.00

Metal stud partition, faced both sides
with plasterboard, one hour fire
resistance

170mm thick	m2	64.00
200mm thick	m2	80.00

Demountable steel framed
Partitions

vinyl faced	m2	170.00
Softwood framed glazed screens	m2	220.00
Hardwood framed glazed screens	m2	278.00

INTERNAL DOORS (including frames and ironmongery)

Softwood flush door 40mm thick size
762 x 1981mm

plywood faced	nr	140.00
sapele faced	nr	150.00

Softwood flush door 40mm thick size
826 x 2040mm

plywood faced	nr	160.00
sapele faced	nr	170.00

£

Softwood flush door 40mm thick size
826 x 2040mm

plywood faced	nr	185.00
sapele faced	nr	215.00

Softwood flush door 40mm thick,
half hour fire resistant, size
826 x 2040mm

plywood faced	nr	190.00
sapele faced	nr	220.00

Purpose made softwood door
44mm thick size 762 x 1981mm

four panelled	nr	240.00

Purpose made door
with one panel open for
glass, size

four panelled	nr	270.00

Purpose made mahogany door
with one panel open for
glass, size

762 x 1981 x 50.mm	nr	280.00
838 x 1981 x 50mm	nr	290.00
762 x 1981 x 63mm	nr	310.00
838 x 1981 x 63mm	nr	320.00

Purpose made mahogany door
with two panels open for
glass, size

762 x 1981 x 50.mm	nr	300.00
838 x 1981 x 50mm	nr	310.00
762 x 1981 x 63mm	nr	330.00
838 x 1981 x 63mm	nr	340.00

£

Purpose made mahogany door
with four panels open for
glass, size

762 x 1981 x 50.mm	nr	320.00
838 x 1981 x 50mm	nr	330.00
762 x 1981 x 63mm	nr	350.00
838 x 1981 x 63mm	nr	360.00

WALL FINISHES

In situ finishings

One coat of plasterboard finish 3mm thick to plasterboard	m2	7.00

Plasterwork to brick or block walls

13mm thick	m2	10.00
19mm thick	m2	13.00

Cement rendering in two coats
to brick or block walls

13mm thick	m2	9.00
19mm thick	m2	11.00
Plasterboard 9.5mm thick and skim coat	m2	20.00
Plasterboard 12.5mm thick and skim coat	m2	22.00
One mist and two coats of emulsion	m2	7.00
One coat undercoat and two coats gloss	m2	18.00

		£
Lining paper	m2	5.00
Vinyl paper		
(PC £4.00/roll)	m2	6.00
(PC £6.00/roll)	m2	7.00
(PC £8.00/roll)	m2	9.00

Board linings

Dry plasterboard lining to walls
for direct decoration with
emulsion paint finish

9.5mm wallboard	m2	20.00
12.5mm wallboard	m2	22.00

Sheet linings on softwood battens
plugged and screwed to walls

plywood 4mm thick	m2	20.00
plywood 6mm thick	m2	24.00
hardboard 3.2mm thick	m2	20.00
hardboard 4.8mm thick	m2	22.00
hardboard 6.0mm thick	m2	24.00
wallboard 9.5mm thick	m2	20.00
wallboard 12.5mm thick	m2	22.00
chipboard 12mm thick	m2	18.00
chipboard 15mm thick	m2	20.00
chipboard 18mm thick	m2	22.00
chipboard 25mm thick	m2	24.00
softwood boarding 13mm thick	m2	20.00
softwood boarding 25mm thick	m2	22.00

£

Tiling

Glazed ceramic wall tiling fixed with
adhesive and grouted

108 x 108 x 4.mm	m2	40.00
152 x 152 x 5.5mm	m2	36.00
200 x 200 x 7mm	m2	32.00

FLOOR FINISHES

Timber flooring

Softwood flooring, butt jointed

19mm thick	m2	22.00
22mm thick	m2	24.00
25mm thick	m2	26.00

Softwood flooring, tongued
and grooved

19mm thick	m2	24.00
22mm thick	m2	26.00
25mm thick	m2	28.00

Hardwood strip flooring, tongued
and grooved, 25mm thick

utile mahogany	m2	64.00
maple	m2	68.00
iroko	m2	58.00

Chipboard flooring, butt
jointed

18mm thick	m2	14.00
22mm thick	m2	16.00

£

Chipboard flooring, tongued
and grooved

18mm thick	m2	18.00
22mm thick	m2	20.00

Plywood flooring, butt
jointed

15mm thick	m2	24.00
18mm thick	m2	30.00

Plywood flooring, tongued
and grooved

15mm thick	m2	28.00
18mm thick	m2	34.00

Screeds/in situ finishings

Cement and sand (1:3) screed

25mm thick	m2	12.00
38mm thick	m2	15.00
50mm thick	m2	18.00
75mm thick	m2	22.00

Latex screed

3mm thick	m2	6.00
5mm thick	m2	8.00

Granolithic screed

25mm thick	m2	16.00
32mm thick	m2	18.00
38mm thick	m2	20.00
50mm thick	m2	22.00

£

Mastic flooring

    black

20mm thick	m2	22.00
25mm thick	m2	25.00

    red

20mm thick	m2	26.00
25mm thick	m2	28.00

Epoxy floor finish, 5mm thick	m2	30.00

Tile flooring

Quarry tiles

    red

150 x 150 x 12.5mm	m2	48.00
200 x 200 x 25mm	m2	54.00

    brown

150 x 150 x 12.5mm	m2	54.00
200 x 200 x 25mm	m2	60.00

Ceramic tiles

100 x 100 x 9mm	m2	36.00
150 x 150 x 12mm	m2	42.00
200 x 200 x 12mm	m2	56.00

Flexible flooring

Rubber floor tiles, 3mm thick

plain finish, black	m2	42.00
studded finish, black	m2	48.00

£

Linoleum sheeting

2.5mm thick	m2	20.00
3.2mm thick	m2	24.00

Linoleum tiling 3.2mm thick	m2	18.00

Thermoplastic tiling, 2mm thick	m2	12.00

Vinyl sheeting

2mm thick	m2	16.00
2.5mm thick	m2	18.00
3mm thick	m2	20.00

Vinyl tiling, 2mm thick	m2	12.00

Cork tiling

3.2mm thick	m2	26.00
4.8mm thick	m2	30.00
6.3mm thick	m2	34.00
8.00mm thick	m2	40.00

Woodblock flooring herringbone
pattern, sanded and wax polished

maple	m2	64.00
oak	m2	80.00
merbau	m2	58.00
iroko	m2	60.00

Wood strip flooring herringbone
pattern, sanded and wax polished

maple	m2	68.00
oak	m2	84.00
merbau	m2	62.00
iroko	m2	64.00

£

Underlay to carpets		
rubber	m2	8.00
felt	m2	6.00
Fitted carpets		
contract quality		
medium duty	m2	18.00
heavy duty	m2	24.00
Carpet tiles		
domestic	m2	18.00
medium duty	m2	24.00
heavy duty	m2	30.00

Skirtings

Softwood painted skirtings	m2	6.00
Hardwood stained skirtings	m2	12.00
Vinyl coved skirtings		
77mm high	m2	5.00
100mm high	m2	6.00

CEILING FINISHES

In situ finishings

Skim coat of plaster to ceilings	m2	8.00
One mist and two coats of emulsion	m2	8.00
One coat Artex sealer and one coat Artex finish		
plastered ceilings	m2	4.00
plasterboard ceilings	m2	4.00

		£
Lining paper	m2	5.00
Vinyl paper		
(PC £4.00/roll)	m2	7.00
(PC £6.00/roll)	m2	8.00
(PC £8.00/roll)	m2	10.00

Board finishes

Plasterboard 9.5mm thick and skim coat	m2	20.00
Plasterboard 12.5mm thick and skim coat	m2	22.00

Suspended ceiling systems

Gyproc M/F suspended ceiling system with 12.7mm thick wallboard	m2	34.00

Suspended ceiling system
with acoustic tiles

300 x 300mm	m2	25.00
600 x 600mm	m2	26.00

**SANITARY FITTINGS/DISPOSAL
INSTALLATIONS**

**Sanitary fittings (complete with
water supply, taps and waste
pipework)**

White

lavatory basin	nr	180.00
WC	nr	280.00
urinal bowl	nr	180.00
shower cubicle	nr	800.00
sink	nr	230.00
bath, steel enamelled	nr	360.00
bath, acrylic	nr	340.00

£

Coloured
lavatory basin	nr	190.00
WC	nr	290.00
bath, acrylic	nr	350.00

Pipework (including all fittings)

Waste pipes

copper
35mm	m	16.00
42mm	m	20.00
54mm	m	24.00

PVC-U
32mm	m	8.00
40mm	m	9.00

Overflow pipes

copper
22mm	m	8.00
28mm	m	10.00

MuPVC
19mm	m	7.00

Traps

Polypropylene
S trap	nr	14.00
P trap	nr	12.00

Soil pipes

cast iron
75mm	m	44.00
100mm	m	50.00

£

PVC-U

110mm	m	26.00
150mm	m	46.00

HOT AND COLD WATER SERVICES

Pipework (including all fittings)

Copper pipes, capillary fittings

15mm	m	14.00
22mm	m	16.00
28mm	m	20.00
35mm	m	32.00
42mm	m	40.00
54mm	m	48.00

Copper pipes, compression fittings

15mm	m	18.00
22mm	m	20.00
28mm	m	24.00
35mm	m	38.00
42mm	m	48.00
54mm	m	56.00

Cisterns and cylinders (including all connections and overflows)

Galvanised iron cold water
cisterns, capacity

36 litres	nr	94.00
54 litres	nr	100.00
68 litres	nr	110.00
86 litres	nr	120.00
114 litres	nr	132.00
154 litres	nr	180.00

£

Polyethylene cold water cisterns,
capacity

68 litres	nr	48.00
114 litres	nr	60.00
182 litres	nr	84.00
227 litres	nr	90.00

Copper cylinders, indirect pattern
capacity

114 litres	nr	160.00
117 litres	nr	176.00
140 litres	nr	204.00
162 litres	nr	248.00

Copper cylinders, direct pattern
capacity

98 litres	nr	190.00
120 litres	nr	210.00
148 litres	nr	234.00
166 litres	nr	257.00

BUILDERS WORK IN CONNECTION WITH SPECIALIST SERVICES

Holes for pipes 55mm
diameter through

half brick wall	nr	10.00
one brick wall	nr	22.00
one and a half brick wall	nr	30.00
blockwork 100mm thick	nr	8.00
blockwork 140mm thick	nr	10.00
blockwork 190mm thick	nr	12.00

£

Holes for pipes 55 to 110mm
diameter through

half brick wall	nr	14.00
one brick wall	nr	26.00
one and a half brick wall	nr	16.00
blockwork 100mm thick	nr	12.00
blockwork 140mm thick	nr	14.00
blockwork 190mm thick	nr	16.00

Chases in walls for pipes up to
55mm diameter

brickwork	m	8.00
brickwork	m	5.00

Chases in walls for pipes up to
55mm to 110mm diameter

brickwork	m	12.00
brickwork	m	8.00

EXTERNAL WORKS

£

Excavation

Excavate topsoil 150mm thick
and deposit on site in spoil heaps

by machine	m2	2.00
by hand	m2	4.00

Excavate topsoil 150mm thick
and spread and level on site

	m2	
by machine		4.00
by hand	m2	16.00

Excavate to reduce levels and
deposit on site in spoil heaps

depth not exceeding 0.25m

by machine	m3	6.00
by hand	m3	18.00

depth not exceeding 1m

by machine	m3	7.00
by hand	m3	20.00

depth not exceeding 2m

by machine	m3	8.00
by hand	m3	22.00

depth not exceeding 4m

by machine	m3	6.00
by hand	m3	24.00

Excavate to reduce levels and
spread and level on site

depth not exceeding 0.25m

by machine	m3	7.00
by hand	m3	20.00

			£

depth not exceeding 1m

by machine	m3		8.00
by hand	m3		24.00

depth not exceeding 2m

by machine	m3		9.00
by hand	m3		26.00

depth not exceeding 4m

by machine	m3		10.00
by hand	m3		28.00

Excavate to reduce levels and
remove to tip off site

depth not exceeding 0.25m

by machine	m3		20.00
by hand	m3		26.00

depth not exceeding 1m

by machine	m3		22.00
by hand	m3		28.00

depth not exceeding 2m

by machine	m3		24.00
by hand	m3		30.00

depth not exceeding 4m

by machine	m3		26.00
by hand	m3		32.00

Breaking up

Excavate in soft rock

by machine	m3		45.00
by hand	m3		65.00

£

Excavate in hard rock

| | by machine | m3 | 54.00 |
| | by hand | m3 | 74.00 |

Excavate in concrete

| | by machine | m3 | 35.00 |
| | by hand | m3 | 55.00 |

Excavate in masonry

| | by machine | m3 | 30.00 |
| | by hand | m3 | 50.00 |

Excavate in pavings 150mm thick

Concrete

| | by machine | m2 | 3.00 |
| | by hand | m2 | 4.00 |

macadam

| | by machine | m2 | 2.00 |
| | by hand | m2 | 4.00 |

Filling

Imported filling material deposited
on site in layers not exceeding 250mm thick, compacting
with vibrating roller

excavating material

| | by machine | m3 | 6.00 |
| | by machine | m3 | 12.00 |

sand

| | by machine | m3 | 35.00 |
| | by hand | m3 | 50.00 |

£

hardcore
by machine	m3	30.00
by hand	m3	45.00

granular fill Type 1
by machine	m3	35.00
by hand	m3	50.00

granular fill Type 2
by machine	m3	33.00
by hand	m3	48.00

Surface treatments

Level and compact

excavation bottom	m2	2.00
filling	m2	3.00

Trim

sloping excavated surfaces	m2	2.00
sloping rock	m2	10.00

Beds to receive pavings

Hardcore in bed

75mm thick
by machine	m2	3.00
by hand	m2	5.00

100mm thick
by machine	m2	3.50
by hand	m2	6.00

£

150mm thick		
by machine	m2	4.00
by hand	m2	6.50

Quarry waste in bed

75mm thick		
by machine	m2	3.50
by hand	m2	5.50
100mm thick		
by machine	m2	4.00
by hand	m2	6.50
150mm thick		
by machine	m2	4.50
by hand	m2	7.00

Granular fill Type 1 in bed
by machine

100mm thick	m2	6.00
150mm thick	m2	7.00
200mm thick	m2	9.00
250mm thick	m2	11.00
300mm thick	m2	14.00

Pavings (laid on prepared bed)

In situ concrete with trowelled
finish

100mm thick	m2	10.00
125mm thick	m2	12.00
150mm thick	m2	14.00

£

Precast concrete paving flags
50mm thick

600 x 450mm			
	natural	m2	20.00
	coloured	m2	22.00
600 x 600mm			
	natural	m2	20.00
	coloured	m2	22.00
600 x 750mm			
	natural	m2	18.00
	coloured	m2	20.00
600 x 900mm			
	natural	m2	18.00
	coloured	m2	20.00

Concrete block paviours size
200 x 100 x 60mm thick, laid

straight bond			
	natural	m2	26.00
	coloured	m2	28.00
herringbone pattern			
	natural	m2	30.00
	coloured	m2	32.00

Concrete block paviours size
200 x 100 x 80mm thick, laid

straight bond			
	natural	m2	28.00
	coloured	m2	30.00
herringbone pattern			
	natural	m2	32.00
	coloured	m2	34.00

£

Reconstructed stone paving flags

size 450 x 450 x 40mm	m2	44.00
size 600 x 450 x 40mm	m2	42.00

Brick paviours size 215 x 102.5 x 65mm (£300 per 1000)

laid flat

straight bond	m2	36.00
herringbone pattern	m2	40.00

laid on edge

straight bond	m2	48.00
herringbone pattern	m2	52.00

Brick paviours size 215 x 102.5 x 65mm (£400 per 1000)

laid flat

straight bond	m2	44.00
herringbone pattern	m2	48.00

laid on edge

straight bond	m2	56.00
herringbone pattern	m2	60.00

Granite setts size 200 x 100 x 100mm	m2	80.00
Cobble paving average size 75mm	m2	94.00

York stone paving

600 x 600 x 50mm thick	m2	110.00
600 x 900 x 50mm thick	m2	112.00

£

Fencing

Fencing (including excavating
for post holes, intermediate and
end posts and concrete bases)

Chestnut fencing to BS1722 Part 4

1.00m high	m	12.00
1.25m high	m	13.00
1.50m high	m	14.00
1.80m high	m	16.00

Chainlink fencing to BS1722
Part 1 with galvanised wire
mesh, height

galvanised mild steel posts

0.90m	m	18.00
1.20m	m	22.00
1.80m	m	26.00

concrete posts

0.90m	m	16.00
1.20m	m	20.00
1.80m	m	24.00

Chainlink fencing to BS1722
Part 1 with plastic coated wire
mesh, height

galvanised mild steel posts

0.90m	m	20.00
1.20m	m	24.00
1.80m	m	28.00

concrete posts

0.90m	m	18.00
1.20m	m	22.00
1.80m	m	26.00

£

Close boarded fencing to BS1772
Part 5

on timber posts		
1.00m high	m	40.00
1.25m high	m	44.00
1.50m high	m	50.00
1.80m high	m	52.00
on concrete posts		
1.00m high	m	46.00
1.25m high	m	50.00
1.50m high	m	56.00

Panel fencing to BS1722 Part 11

on timber posts		
0.90m high	m	22.00
1.20m high	m	28.00
1.50m high	m	32.00
1.80m high	m	34.00
on concrete posts		
0.90m high	m	24.00
1.20m high	m	30.00
1.50m high	m	34.00
1.80m high	m	36.00

Palisade fencing to BS1722 Part 6

on timber posts		
1.90m high	m	24.00
1.20m high	m	26.00
1.50m high	m	28.00
1.80m high	m	30.00
on concrete posts		
0.90m high	m	26.00
1.20m high	m	28.00
1.50m high	m	30.00
1.80m high	m	32.00

£

Post and rail fencing to BS1722
Part 7

three rail morticed		
1.10m high	m	24.00
three rail nailed		
1.10m high	m	20.00
four rail morticed		
1.10m high	m	26.00
1.30m high	m	28.00
four rail nailed		
1.10m high	m	22.00
1.30m high	m	24.00

Drainage

Excavate trench by machine,
lay 100mm vitrified clay push fit
flexible pipe with

granular filling to bed and haunching		
depth 0.50m	m	30.00
depth 0.75m	m	34.00
depth 1.00m	m	38.00
depth 1.25m	m	46.00
depth 1.50m	m	50.00
depth 1.75m	m	56.00
depth 2.00m	m	66.00
depth 2.25m	m	74.00
depth 2.50m	m	80.00
depth 2.75m	m	92.00
depth 3.00m	m	98.00
granular filling to bed and surround		
depth 0.50m	m	32.00
depth 0.75m	m	36.00
depth 1.00m	m	40.00

		£
depth 1.25m	m	48.00
depth 1.50m	m	52.00
depth 1.75m	m	58.00
depth 2.00m	m	68.00
depth 2.25m	m	76.00
depth 2.50m	m	82.00
depth 2.75m	m	94.00
depth 3.00m	m	100.00

concrete to bed and
haunching

depth 0.50m	m	36.00
depth 0.75m	m	40.00
depth 1.00m	m	44.00
depth 1.25m	m	52.00
depth 1.50m	m	56.00
depth 1.75m	m	62.00
depth 2.00m	m	72.00
depth 2.25m	m	80.00
depth 2.50m	m	86.00
depth 2.75m	m	98.00
depth 3.00m	m	104.00

concrete to bed and
surround

depth 0.50m	m	41.00
depth 0.75m	m	45.00
depth 1.00m	m	49.00
depth 1.25m	m	57.00
depth 1.50m	m	61.00
depth 1.75m	m	67.00
depth 2.00m	m	77.00
depth 2.25m	m	85.00
depth 2.50m	m	91.00
depth 2.75m	m	103.00
depth 3.00m	m	109.00

£

Excavate trench by machine,
lay 150mm vitrified clay push fit
flexible pipe with

granular filling to
bed and haunching

depth 0.50m	m	36.00
depth 0.75m	m	40.00
depth 1.00m	m	44.00
depth 1.25m	m	52.00
depth 1.50m	m	56.00
depth 1.75m	m	62.00
depth 2.00m	m	72.00
depth 2.25m	m	80.00
depth 2.50m	m	86.00
depth 2.75m	m	98.00
depth 3.00m	m	108.00

granular filling to
bed and surround

depth 0.50m	m	38.00
depth 0.75m	m	42.00
depth 1.00m	m	46.00
depth 1.25m	m	54.00
depth 1.50m	m	58.00
depth 1.75m	m	64.00
depth 2.00m	m	74.00
depth 2.25m	m	82.00
depth 2.50m	m	88.00
depth 2.75m	m	100.00
depth 3.00m	m	106.00

concrete to bed and
haunching

depth 0.50m	m	42.00
depth 0.75m	m	46.00
depth 1.00m	m	50.00
depth 1.25m	m	58.00
depth 1.50m	m	62.00
depth 1.75m	m	68.00
depth 2.00m	m	78.00
depth 2.25m	m	86.00

		£
depth 2.50m	m	92.00
depth 2.75m	m	100.00
depth 3.00m	m	106.00

concrete to bed and
surround

depth 0.50m	m	47.00
depth 0.75m	m	51.00
depth 1.00m	m	55.00
depth 1.25m	m	63.00
depth 1.50m	m	67.00
depth 1.75m	m	73.00
depth 2.00m	m	83.00
depth 2.25m	m	91.00
depth 2.50m	m	97.00
depth 2.75m	m	109.00
depth 3.00m	m	115.00

Excavate trench by hand, lay
100mm vitrified clay push fit
flexible pipe with

granular filling to
bed and haunching

depth 0.50m	m	35.00
depth 0.75m	m	42.00
depth 1.00m	m	47.00
depth 1.25m	m	60.00
depth 1.50m	m	70.00
depth 1.75m	m	85.00
depth 2.00m	m	100.00
depth 2.25m	m	116.00
depth 2.50m	m	136.00
depth 2.75m	m	152.00
depth 3.00m	m	170.00

granular filling to
bed and surround

depth 0.50m	m	37.00
depth 0.75m	m	44.00
depth 1.00m	m	49.00
depth 1.25m	m	62.00

		£
depth 1.50m	m	72.00
depth 1.75m	m	87.00
depth 2.00m	m	102.00
depth 2.25m	m	118.00
depth 2.50m	m	133.00
depth 2.75m	m	150.00
depth 3.00m	m	160.00

concrete to bed and
haunching

depth 0.50m	m	41.00
depth 0.75m	m	48.00
depth 1.00m	m	53.00
depth 1.25m	m	66.00
depth 1.50m	m	76.00
depth 1.75m	m	91.00
depth 2.00m	m	106.00
depth 2.25m	m	122.00
depth 2.50m	m	137.00
depth 2.75m	m	154.00
depth 3.00m	m	164.00

concrete to bed and
surround

depth 0.50m	m	46.00
depth 0.75m	m	53.00
depth 1.00m	m	58.00
depth 1.25m	m	71.00
depth 1.50m	m	81.00
depth 1.75m	m	96.00
depth 2.00m	m	111.00
depth 2.25m	m	127.00
depth 2.50m	m	142.00
depth 2.75m	m	159.00
depth 3.00m	m	169.00

£

Manholes

Brick manholes including
excavation, concrete base,
engineering brick walls,
main channel and bends,
benching and cast iron
inspection cover, internal size

600 x 500mm			
depth 500mm	m		350.00
depth 750mm	m		380.00
depth 1000mm	m		435.00
depth 1250mm	m		495.00
depth 1500mm	m		580.00
750 x 450mm			
depth 500mm	m		375.00
depth 750mm	m		400.00
depth 1000mm	m		445.00
depth 1250mm	m		555.00
depth 1500mm	m		620.00
900 x 600mm			
depth 500mm	m		425.00
depth 750mm	m		540.00
depth 1000mm	m		625.00
depth 1250mm	m		740.00
depth 1500mm	m		860.00
depth 1750mm	m		980.00

£

Brick manholes including
excavation, concrete base,
engineering brick walls,
main channel and bends,
benching, concrete reducing
slab and cast iron inspection
cover, internal size

900 x 900mm

depth 1500mm	m	1100.00
depth 2000mm	m	1200.00
depth 2500mm	m	1350.00
depth 3000mm	m	1500.00
depth 3500mm	m	1650.00
depth 4000mm	m	1800.00

1200 x 750mm

depth 1500mm	m	900.00
depth 2000mm	m	1100.00
depth 2500mm	m	1300.00
depth 3000mm	m	1500.00
depth 3500mm	m	1700.00
depth 4000mm	m	1900.00

900 x 1500mm

depth 1500mm	m	1300.00
depth 2000mm	m	1400.00
depth 2500mm	m	1600.00
depth 3000mm	m	1750.00
depth 3500mm	m	2200.00
depth 4000mm	m	2500.00

1200 x 1800mm

depth 1500mm	m	1400.00
depth 2000mm	m	1500.00
depth 2500mm	m	1600.00
depth 3000mm	m	1850.00
depth 3500mm	m	2375.00
depth 4000mm	m	2700.00

LANDSCAPING

£

Site clearance

Demolish existing buildings
Including digging up foundations

brick		
small	m3	12.00
medium	m3	9.00
large	m3	6.00
steel framed with cladded walls		
small	m3	5.00
medium	m3	4.00
large	m3	3.00
timber framed with cladded walls		
small	m3	3.00
medium	m3	2.00
large	m3	1.00

Temporary chestnut fencing
1.50m high   m   14.00

Clear away scrub vegetation,
shrub and hedges   m2   3.00

Cut down trees, grub up roots

trees size less than 600mm girth	nr	70.00
trees size 600 to 900mm girth	nr	150.00
trees size 900 to 1200mm girth	nr	300.00
trees size 1200 to 1500mm girth	nr	350.00
trees size 1500 to 1800mm girth	nr	450.00
trees size 1800 to 2100mm girth	nr	550.00
trees size 2100 to 2400mm girth	nr	650.00
trees size 2400 to 2700mm girth	nr	750.00
trees size 2700 to 3000mm girth	nr	850.00

£

Backfill tree hole with excavated
material

trees size less than 600mm girth	nr	3.00
trees size 600 to 900mm girth	nr	4.00
trees size 900 to 1200mm girth	nr	5.00
trees size 1200 to 1500mm girth	nr	6.00
trees size 1500 to 1800mm girth	nr	8.00
trees size 1800 to 2100mm girth	nr	10.00
trees size 2100 to 2400mm girth	nr	12.00
trees size 2400 to 2700mm girth	nr	16.00
trees size 2700 to 3000mm girth	nr	20.00

Backfill tree hole with sand

trees size less than 600mm girth	nr	12.00
trees size 600 to 900mm girth	nr	16.00
trees size 900 to 1200mm girth	nr	24.00
trees size 1200 to 1500mm girth	nr	36.00
trees size 1500 to 1800mm girth	nr	44.00
trees size 1800 to 2100mm girth	nr	50.00
trees size 2100 to 2400mm girth	nr	60.00
trees size 2400 to 2700mm girth	nr	70.00
trees size 2700 to 3000mm girth	nr	80.00

Excavation and filling

Excavate topsoil 150mm thick
and deposit on site in spoil heaps

by machine	m2	2.00
by hand	m2	4.00

Excavate topsoil 150mm thick
and spread and level on site

by machine	m2	4.00
by hand	m2	16.00

£

Imported filling material
deposited on site in layers not
exceeding 250mm thick, compacting
with vibrating roller

    excavated material

	by machine	m2	6.00
	by hand	m2	12.00

    sand

	by machine	m2	35.00
	by hand	m2	50.00

    hardcore

	by machine	m2	30.00
	by hand	m2	45.00

    granular fill Type 1

	by machine	m2	35.00
	by hand	m2	50.00

    granular fill Type 2

	by machine	m2	33.00
	by hand	m2	48.00

Surface treatments

Break up existing ground
with plough or rotovator,
depth

100mm	m2	1.50
200mm	m2	1.60
300mm	m2	1.75
400mm	m2	2.00

		£
Roll cultivated ground with self-propelled roller	m2	0.75
Level and compact		
excavation bottom	m2	1.50
filling	m2	2.00
Trim surfaces of		
sloping excavated surfaces	m2	2.00
sloping rock	m2	10.00

Soil stabilisation

Biogradable unseeded erosion control mats 2400mm wide fixed with pins to prepared ground

Eromat Light	m2	5.00
Eromat Standard	m2	5.50
Eromat Coco	m2	6.00

Biogradable seeded erosion control mats 2400mm wide fixed with pins to prepared ground

Covamat Standard	m2	6.00
Covamat Special	m2	6.50
Covamat Coco	m2	6.50

Grassblock precast concrete paving, laid on sand bed, filled in with topsoil and seeded

83mm thick	m2	24.00
103mm thick	m2	27.00
125mm thick	m2	30.00

£

Grasscrete polystyrene
formers, concrete infilling,
topsoil and seeding

100mm thick	m2	28.00
150mm thick	m2	32.00

Retaining walls

Excavate for and place
in position wire mesh
gabion cages filled with
broken stone and rock, size

zinc mesh

2 x 1 x 0.5m	nr	100.00
2 x 1 x 1m	nr	115.00

PVC coated wire mesh

2 x 1 x 0.5m	nr	110.00
2 x 1 x 1m	nr	125.00

Excavate for and place
in position galvanised wire
mesh gabion mattresses
filled with broken stone
and rock, size

6 x 2 x 0.25m	nr	28.00
6 x 2 x 0.30m	nr	32.00

Excavate trench by machine, dispose of
surplus excavated material off site,
earthwork support, concrete foundations
10 N/mm2 – 40mm aggregate (1:3:6),
engineering bricks (£200/1000) in
retaining wall

one brick thick wall

180 wall height 1500mm	m	210.00
250 wall height 2000mm	m	280.00
330 wall height 2000mm	m	360.00

			£
one and a half brick thick wall			
240 wall height 1500mm	m		270.00
320 wall height 2000mm	m		350.00
400 wall height 2000mm	m		430.00

Roads

Excavate for and lay road base,
tarmacadam sub-base and wearing
course, precast concrete kerbs both
sides, road width

5m	m		250.00
6m	m		450.00
7m	m		650.00

Paths

Excavate for hardcore base,
blinded with sand, covered in
bark chippings laid between
softwood edging boards

width, 1.5m	m		30.00
width, 2.5m	m		40.00
width, 2.5m	m		50.00
width, 3.0m	m		60.00

Excavate for hardcore base,
blinded with sand, lay precast
concrete blocks as fire path
infilled with top soil and seeded

width, 3.0m	m		96.00
width, 4.0m	m		128.00
width, 5.0m	m		260.00

Precast concrete flags size
600 x 600 x 50mm laid
separately as stepping stones
including excavation and
sand bed nr 6.00

£

Car parks

Excavate and lay sub-base
to receive

tarmacadam 65mm thick		
general area	m2	30.00
parking bay	nr	540.00
concrete blocks 65mm thick		
general area	m2	35.00
parking bay	nr	630.00

cellular precast concrete
paving, filled in with topsoil
and seeded

83mm thick		
general area	m2	24.00
parking bay	nr	432.00
103mm thick		
general area	m2	27.00
parking bay	nr	486.00
125mm thick		
general area	m2	30.00
parking bay	nr	540.00

Sports grounds

Trim and grade prepared
ground and apply weedkiller,
fertiliser, grass seed including
harrowing, rolling and one cut

general areas	m2	0.45
football pitch	nr	4,670.00
rugby pitch	nr	3,100.00
hockey pitch	nr	2,250.00

£

Soiling, seeding and turfing

Imported topsoil filling spread
and levelled, average thickness

by machine		
100mm	m2	2.20
150mm	m2	2.40
200mm	m2	2.60
250mm	m2	2.80
by hand		
100mm	m2	2.80
150mm	m2	4.20
200mm	m2	5.50
250mm	m2	7.00

Topsoil filling from spoil
heaps on site, spread and
levelled, average thickness

by machine		
100mm	m2	0.20
150mm	m2	0.40
200mm	m2	0.60
250mm	m2	0.80
by hand		
100mm	m2	0.80
150mm	m2	1.20
200mm	m2	1.50
250mm	m2	2.00

Plough and harrow topsoil
to fine tilth, remove stones
apply weed killer | m2 | 0.25

£

Apply pesticide and weedkiller
to prepared topsoil, sow grass
seed (£80 per 25kg), harrow, roll
and one cut

by hand			
	10g/m2	m2	0.25
	15g/m2	m2	0.27
	20g/m2	m2	0.28
	25g/m2	m2	0.30
	30g/m2	m2	0.32
	35g/m2	m2	0.34
	40g/m2	m2	0.35
	50g/m2	m2	0.40
by machine			
	10g/m2	m2	0.15
	15g/m2	m2	0.16
	20g/m2	m2	0.17
	25g/m2	m2	0.18
	30g/m2	m2	0.20
	35g/m2	m2	0.22
	40g/m2	m2	0.23
	50g/m2	m2	0.25

Imported turf size 4000 x 2000
x 19mm on prepared bed

general sports use	m2	3.00
special sports use	m2	4.00
domestic	m2	4.50

Treat turf with wooden
paddle beater — m2 — 0.30

Treat turf with light roller — m2 — 0.50

First cut to turf 20mm high
with man-operated power
driven cylinder mower
including boxing cuttings — m2 — 0.04

£

Planting

Excavate for and plant
transplants or seedlings,
backfill and water,
plant cost

£0.50	nr	1.25
£1.00	nr	1.75
£1.50	nr	2.25
£2.00	nr	2.75
£2.50	nr	3.25
£3.00	nr	1.75

Excavate tree pit, fork
bottom, plant tree, backfill
water , surround with peat
(1 stake and 2 ties per tree),
tree cost

£5.00	nr	45.00
£10.00	nr	50.00
£15.00	nr	55.00
£20.00	nr	60.00
£50.00	nr	90.00
£75.50	nr	115.00
£100.00	nr	140.00

Form planting hole in cultivated
area, plant shrub, backfill, water,
surround with peat, shrub cost

£5.00	nr	10.00
£10.00	nr	20.00

Tree stake 100mm diameter and
two ties

length, 2m	nr	10.00
length, 2.5m	nr	11.00
length, 3m	nr	12.00

£

Galvanised wire tree guards,
300mm diameter

height, 1m	nr	20.00
height, 1.5m	nr	25.00
height, 2m	nr	30.00

Cast iron tree grille, size

two part

1000 x 1000mm	nr	200.00
1200 x 1200mm	nr	240.00
1200mm diameter	nr	210.00

four part

1000 x 1000mm	nr	250.00
1200 x 1200mm	nr	300.00
1200mm diameter	nr	280.00

Field drains

Excavate by hand vee-sided ditch,
width at bottom

300mm, depth

750mm	m	12.00
1000mm	m	16.00

500mm, depth

750mm	m	16.00
1000mm	m	20.00
1250mm	m	22.00
1500mm	m	24.00

Excavate trench by machine
and lay agricultural clay pipes,
backfill with gravel rejects

75mm diameter

depth, 400mm	m	15.00
depth, 500mm	m	16.50
depth, 600mm	m	17.50

£

depth, 700mm	m	18.50
depth, 800mm	m	19.50
depth, 900mm	m	20.50
depth, 1000mm	m	21.50

100mm diameter

depth, 400mm	m	18.00
depth, 500mm	m	19.50
depth, 600mm	m	20.50
depth, 700mm	m	21.50
depth, 800mm	m	22.50
depth, 900mm	m	23.50
depth, 1000mm	m	24.50

150mm diameter

depth, 400mm	m	22.00
depth, 500mm	m	23.50
depth, 600mm	m	24.50
depth, 700mm	m	25.50
depth, 800mm	m	26.50
depth, 900mm	m	27.50
depth, 1000mm	m	28.50

Excavate trench by hand
and lay agricultural clay pipes,

75mm diameter

depth, 400mm	m	21.00
depth. 500mm	m	22.00
depth, 600mm	m	24.50
depth, 700mm	m	26.00
depth, 800mm	m	27.50
depth, 900mm	m	29.00
depth, 1000mm	m	30.50

100mm diameter

depth, 400mm	m	24.00
depth, 500mm	m	26.00
depth, 600mm	m	27.50
depth, 700mm	m	29.00
depth, 800mm	m	30.50

		£
depth, 900mm	m	32.00
depth, 1000mm	m	33.50
150mm diameter		
depth, 400mm	m	28.00
depth, 500mm	m	30.00
depth, 600mm	m	31.50
depth, 700mm	m	33.00
depth, 800mm	m	34.50
depth, 900mm	m	36.00
depth, 1000mm	m	37.50

CIVIL ENGINEERING WORK

Individual rates in civil engineering work cannot be compared to those in building because the value of the General Items can be as high as 50% compared to about 10% in building work. This is mainly due to the freedom of civil engineering estimators to prepare bids to reflect their view of the timing and sequence of the works that is usually not available to their building counterparts.

So when preparing first stage estimates in civil engineering, it is important that the level of the General Items is assessed first so that the level of pricing of the individual rates can be adjusted accordingly.

CLASS A: GENERAL ITEMS

The items listed below are based on a civil contract worth approximately £10m and a contract period of 78 weeks.

Contractual requirements	£	£
Performance bond		
1% for construction period		
x £10m	150,000	
0.75% for maintenance period		
x £10m	75,000	225,000
Insurance of the Works		
1.5% x £10m		150,000
Insurance of the plant		
included in hire charges		nil
Insurance against damage to persons		
and property		
included in head office overheads		nil
Specified requirements		
Offices for Engineer's staff		
erect	1,000	
Carried forward	1,000	375,000

	£	£
Brought forward	1,000	375,000
maintain and operate (100 weeks x £100)	10,000	
remove	1,000	12,000
Laboratory for the Engineer's staff		
erect	750	
maintain and operate (50 weeks x £100)	5,000	
remove	750	6,500
Cabins for the Engineer's staff		
erect	1,000	
maintain and operate (100 weeks x £100)	10,000	
remove	1,000	12,000
Services for the Engineer		
1,750cc car (100 weeks x £300)	30,000	
Land Rover (78 weeks x £300)	23,400	
telephone installation	300	
maintain and operate (100 weeks x £50)	5,000	58,700
Equipment for the Engineer		
office equipment desks, computer, tables, chairs, filing cabinets, sundries (100 weeks x £150)	15,000	
laboratory equipment (50 weeks x £150)	7,500	
surveying equipment (78 weeks x £100)	7,800	30,300
Carried forward		494,500

	£	£
Brought forward		494,500
Attendance on the Engineer		
driver		
(50 weeks x £400)	20,000	
chainman		
(50 weeks x £300)	15,000	
laboratory assistant		
(30 weeks x £400)	<u>12,000</u>	49,000
Testing of materials (included)		nil
Testing of the Works (included)		nil
Temporary works		
traffic signals		
(30 weeks x £50)	1,500	
cleaning roads		
(40 weeks x £400)	16,000	
progress photography	1,000	
temporary lighting		
(40 weeks x £60)	2,400	
temporary water supply		
connection	1,500	
pipework		
(200m x £10)	2,000	
supply		
(2m litres x £0.60 per k)	1,200	
hardstanding		
(lay 200m2 x £12)	2,400	
remove	<u>1,000</u>	29,000

Method related charges

Offices for contractor		
erect	1,000	
maintain and operate		
(100 weeks x £100)	10,000	
remove	<u>1,000</u>	<u>12,000</u>
Carried forward		584,500

	£	£
Brought forward		584,000
Cabins for contractor		
erect	1,000	
maintain and operate		
(78 weeks x £60)	4,680	
remove	1,000	6,680
Stores for contractor		
erect	1,000	
maintain and operate		
(100 weeks x £60)	6,000	
remove	1,000	8,000
Canteen and messroom for contractor		
erect	1,000	
maintain and operate		
(78 weeks x £80)	6,240	
remove	1,000	8,240
Electricity		
install	1,000	
maintain and operate		
(100 weeks x £50)	5,000	6,000
Supervision/administration		
agent		
(90 weeks x £1,000)	90,000	
under agents (2)		
(156 weeks x £600)	93,600	
inspectors (2)		
(156 weeks x £500)	78,000	
setting out engineeer		
(30 weeks x £600)	18,000	
quantity surveyor		
(110 weeks x £800)	88,000	
section foremen (4)		
(312 weeks x £500)	156,000	
timekeeper/wages clerk		
(78 weeks x £400)	31,200	554,800
Carried forward		1,167,720

	£	£
Brought forward		1,167,720
storekeeper		
(78 weeks x £400)	31,200	
watchman		
(100 weeks x £350)	35,000	
teaboy		
(78 weeks x £300)	23,400	
offloading and cleaning		
gang (2)		
(140 weeks x £350)	49,000	138,600
Total		1,444.920

CLASS B: GROUND INVESTIGATION   £

Trial pits

Trial pits size 1 x 2m, not in rock,
maximum depth

not exceeding 1m	nr	24.00
1-2m	nr	38.00
2-3m	nr	50.00
3-5m	nr	60.00

Trial pits size 1 x 2m, partly in rock,
maximum depth

not exceeding 1m	nr	32.00
1-2m	nr	46.00
2-3m	nr	60.00
3-5m	nr	70.00

£

Trial pits size 1 x 2m, in rock,
maximum depth

not exceeding 1m	nr	50.00
1-2m	nr	200.00
2-3m	nr	300.00
3-5m	nr	500.00

An allowance of £2,000 should be made for the establishment
and removal of plant and equipment to carry out the
following work.

Light percussion boreholes, 150mm
diameter

depth, not exceeding 5m	m	25.00
depth, 5-10m	m	30.00
depth, 10-20	m	40.00
depth, 20-30m	m	60.00
depth, 30-40m	m	100.00

Rotary drilled boreholes, 150mm
diameter without core recovery

depth, not exceeding 5m	m	30.00
depth, 5-10m	m	30.00
depth, 10-20	m	30.00
depth, 20-30m	m	30.00
depth, 30-40m	m	30.00

Rotary drilled boreholes, 150mm
diameter with core recovery

depth, not exceeding 5m	m	70.00
depth, 5-10m	m	80.00
depth, 10-20	m	90.00
depth, 20-30m	m	100.00
depth, 30-40m	m	110.00

CLASS C: GEOTECHNICAL AND OTHER SPECIAL PROCESSES

£

Drilling

An allowance of £5,000 should be made for the establishment and removal of plant and equipment to carry out the following work.

Drilling for grout holes through materials other than rock or artificial hard materials

vertically downwards		
depth, not exceeding 5m	m	28.00
depth, 5-10m	m	30.00
depth, 10-20	m	32.00
depth, 20-30m	m	34.00
downwards at an angle of 0-45 degrees to the vertical		
depth, not exceeding 5m	m	28.00
depth, 5-10m	m	30.00
depth, 10-20	m	32.00
depth, 20-30m	m	34.00
horizontally or downwards at an angle of 0-45 degrees to the horizontal		
depth, not exceeding 5m	m	28.00
depth, 5-10m	m	30.00
depth, 10-20	m	32.00
depth, 20-30m	m	34.00
upwards at an angle of 0-45 degrees to the horizontal		
depth, not exceeding 5m	m	34.00
depth, 5-10m	m	36.00
depth, 10-20	m	38.00
depth, 20-30m	m	40.00

		£
upwards at an angle of 0-45 degrees to the vertical		
depth, not exceeding 5m	m	34.00
depth, 5-10m	m	36.00
depth, 10-20	m	38.00
depth, 20-30m	m	40.00

Drilling for grout holes through rock

vertically downwards		
depth, not exceeding 5m	m	26.00
depth, 5-10m	m	28.00
depth, 10-20m	m	30.00
depth, 20-30m	m	32.00

downwards at an angle of 0-45 degrees to the vertical		
depth, not exceeding 5m	m	26.00
depth, 5-10m	m	28.00
depth, 10-20m	m	30.00
depth, 20-30m	m	32.00

horizontally or downwards at an angle of 0-45 degrees to the horizontal		
depth, not exceeding 5m	m	26.00
depth, 5-10m	m	28.00
depth, 10-20m	m	30.00
depth, 20-30m	m	32.00

upwards at an angle of 0-45 degrees to the horizontal		
depth, not exceeding 5m	m	32.00
depth, 5-10m	m	34.00
depth, 10-20m	m	36.00
depth, 20-30m	m	38.00

upwards at an angle of 0-45 degrees to the vertical		
depth, not exceeding 5m	m	32.00
depth, 5-10m	m	34.00
depth, 10-20m	m	36.00
depth, 20-30m	m	38.00

£

Grout materials and injections

An allowance of £3,000 should be made for
the establishment and renewal of plant and
equipment to carry out grouting operations.

Materials

Cement bentonite (2:1)	t	120.00
Cemant PFA	t	80.00
Sand	t	20.00
Pea gravel	t	20.00
Bentonite	t	180.00

Diaphragm walls

It is assumed in the following rates that a minimum of 3,000 m3
of excavation is required. An allowance of £60,000 should be
made for the establishment and removal of plant and equipment
to carry out the work.

Excavation in material other
than rock or artificial hard
material

maximum depth not exceeding 5m	m3	300.00
maximum depth 5-10m	m3	320.00
maximum depth 10-15m	m3	340.00
maximum depth 15-20m	m3	360.00

£

Concrete designed mix to
BS5328; grade 20, ordinary
Portland cement to BS12, 20mm
aggregate to BS882,
walls 1000mm thick | m3 | 130.00

High yield bar reinforcement
To BS4449

nominal size 12mm	t	650.00
nominal size 16mm	t	630.00
nominal size 20mm	t	600.00
nominal size 25mm	t	580.00

Concrete guide wall at the
side of excavation, 1000mm
wide x 500mm deep | m | 275.00

Ground anchorages

It is assumed in the following rates that a minimum of
75 anchors are to be installed and an allowance of £6,000
should be made for the establishment and removal of plant
and equipment.

Ground anchorages, in material
other than rock to a maximum
10m depth; load 50 tonne

temporary	nr	100.00
temporary with single corrosion protection	nr	120.00
temporary with double corrosion protection	nr	140.00
permanent	nr	100.00
permanent with single corrosion protection	nr	120.00
permanent with double corrosion protection	nr	140.00

£

Total length of tendon in material
other than rock

temporary	nr	80.00
temporary with single corrosion protection	nr	85.00
temporary with double corrosion protection	nr	90.00
permanent	nr	95.00
permanent with single corrosion protection	nr	100.00
permanent with double corrosion protection	nr	105.00

Ground anchorages, in material
including rock to a maximum
10m depth; load 50 tonne

temporary	nr	125.00
temporary with single corrosion protection	nr	145.00
temporary with double corrosion protection	nr	165.00
permanent	nr	125.00
permanent with single corrosion protection	nr	145.00
permanent with double corrosion protection	nr	165.00

Total length of tendon in material
including rock

temporary	nr	90.00
temporary with single corrosion protection	nr	95.00
temporary with double corrosion protection	nr	100.00
permanent	nr	105.00
permanent with single corrosion protection	nr	110.00
permanent with double corrosion protection	nr	115.00

£

Sand, band and wick drains

It is assumed in the following rates that a minimum
of 100 vertical drains are to be installed and an
allowance of £7,500 should be made for the
establishment and removal of plant and equipment.

Number of drains	70 nr		
Pre-drilled holes	70 nr		

Drains of maximum depth not
exceeding 10m

cross section		
100-200mm	m	8.00
200-300mm	m	9.00
300-400mm	m	10.00

Drains of maximum depth not
exceeding 10-15m

cross section		
100-200mm	m	9.00
200-300mm	m	10.00
300-400mm	m	11.00

Drains of maximum depth not
exceeding 10-15m

cross section		
100-200mm	m	10.00
200-300mm	m	11.00
300-400mm	m	12.00

CLASS D: DEMOLITION AND SITE CLEARANCE

£

General clearance

General site clearance of areas

free from major obstructions	ha	900.00
woods, small trees and shrubs	ha	1,800.00

Demolish existing buildings including
digging up foundations

brick
small	m3	10.00
medium	m3	8.00
large	m3	6.00

steel framed with cladded walls
small	m3	5.00
medium	m3	4.00
large	m3	3.00

timber framed with cladded walls
small	m3	4.00
medium	m3	3.00
large	m3	2.00

Pull down trees (stumps
measured separately) girth

500mm-1m	nr	60.00
1-2m	nr	100.00
2-3	nr	400.00
3-5m	nr	900.00

£

Grub up stumps and backfill
with displaced topsoil, stump
diameter

150-500mm	nr	30.00
500mm-1	nr	60.00
1-2m	nr	80.00

Dig out and remove drains
depth 1.5m including granular
bed and surround   m   10.00

Dig out and remove drains depth
2m including concrete bed
and surround   m   13.00

CLASS E – EARTHWORKS

Dredging

It is difficult to provide accurate cost information on dredging. Various methods can be used including cutter-suction dredger, barge-mounted excavator or grab hopper. The depth of water, disposal arrangements and tidal conditions are all key factors affecting costs.

The cost per cubic metre of dredging solid material should be in the range of £6 to £8 per cubic metre but specialist advice should be obtained in the early stages of preparing the budget estimate.

General excavation

Excavate to reduce levels

depth not exceeding 0.25m

topsoil	m3	2.00
normal ground	m3	2.00
stiff clay	m3	3.50
chalk	m3	4.50

		£

depth 0.25 to 0.50m

normal ground	m3	2.00
stiff clay	m3	3.50
chalk	m3	20.00
rock	m3	30.00

depth 0.25 to 1m

normal ground	m3	2.00
stiff clay	m3	3.50
chalk	m3	25.00
rock	m3	40.00

depth 1m to 2m

normal ground	m3	2.50
stiff clay	m3	4.00
chalk	m3	30.00
rock	m3	45.00

depth 2 to 5m

normal ground	m3	5.00
stiff clay	m3	8.00
chalk	m3	35.00
rock	m3	50.00

Disposal

Excavated material

deposited on site 100m distance	m3	3.00
deposited on site 300m distance	m3	3.50
deposited off site 1km distance including tipping fees	m3	14.00
deposited off site 5km distance including tipping fees	m3	8.00

£

Filling

Filling to make up levels including
levelling and compacting

excavated material	m3	6.00
sand	m3	28.00
hardcore	m3	22.00
DTp Type 1	m3	26.00
DTp Type 2	m3	25.00

CLASS F: IN SITU CONCRETE

Provision of concrete

Standard mix

ST1

ordinary Portland cement	m3	75.00
sulphate resisting cement	m3	85.00

ST2

ordinary Portland cement	m3	77.00
sulphate resisting cement	m3	87.00

ST3

ordinary Portland cement	m3	80.00
sulphate resisting cement	m3	90.00

ST4

ordinary Portland cement	m3	92.00
sulphate resisting cement	m3	102.00

ST5

ordinary Portland cement	m3	94.00
sulphate resisting cement	m3	104.00

		£

Designed mix, ordinary
Portland cement

grade 20

10mm aggregate	m3	64.00
14mm aggregate	m3	66.00
20mm aggregate	m3	68.00
40mm aggregate	m3	70.00

grade 25

10mm aggregate	m3	66.00
14mm aggregate	m3	68.00
20mm aggregate	m3	70.00
40mm aggregate	m3	72.00

grade 30

10mm aggregate	m3	68.00
14mm aggregate	m3	70.00
20mm aggregate	m3	72.00
40mm aggregate	m3	74.00

Designed mix, sulphate
cement

grade 20

10mm aggregate	m3	74.00
14mm aggregate	m3	76.00
20mm aggregate	m3	78.00
40mm aggregate	m3	80.00

grade 25

10mm aggregate	m3	76.00
14mm aggregate	m3	78.00
20mm aggregate	m3	80.00
40mm aggregate	m3	82.00

grade 30

10mm aggregate	m3	78.00
14mm aggregate	m3	80.00
20mm aggregate	m3	82.00
40mm aggregate	m3	84.00

£

Placing of concrete

Mass concrete in

     blinding, thickness

not exceeding 150mm	m3	16.00
150-300mm	m3	15.00
300-500mm	m3	14.00
exceeding 500mm	m3	13.00

     bases, footings, pile caps
     and ground slabs

not exceeding 150mm	m3	22.00
150-300mm	m3	20.00
300-500mm	m3	18.00
exceeding 500mm	m3	16.00

  walls

not exceeding 150mm	m3	24.00
150-300mm	m3	22.00
300-500mm	m3	20.00
exceeding 500mm	m3	18.00

Reinforced concrete in

     bases, footings, pile caps
     and ground slabs

not exceeding 150mm	m3	24.00
150-300mm	m3	22.00
300-500mm	m3	20.00
exceeding 500mm	m3	18.00

  walls

not exceeding 150mm	m3	26.00
150-300mm	m3	24.00
300-500mm	m3	22.00
exceeding 500mm	m3	20.00

		£
suspended slabs		
not exceeding 150mm	m3	30.00
150-300mm	m3	28.00
300-500mm	m3	26.00
exceeding 500mm	m3	24.00
columns and piers		
not exceeding 150mm	m3	60.00
150-300mm	m3	58.00
300-500mm	m3	56.00
exceeding 500mm	m3	54.00

CLASS G: CONCRETE ANCILLARIES

Formwork

Rough finish

width 0.2m to 0.4m		
horizontal	m2	38.00
vertical	m2	56.00
curved to radius in one plane	m2	86.00
width exceeding 1.22m		
horizontal	m2	34.00
vertical	m2	52.00
curved to radius in one plane	m2	60.00

Fair finish

width 0.2m to 0.4m		
horizontal	m2	40.00
vertical	m2	58.00
curved to radius in one plane	m2	88.00
width exceeding 1.22m		
horizontal	m2	36.00
vertical	m2	54.00
curved to radius in one plane	m2	66.00

£

Reinforcement

Mild steel bars, diameter

6mm	t	1,200
8mm	t	1,180
10mm	t	1,150
12mm	t	1,120
16mm	t	1,100
20mm	t	950
25mm	t	930
32mm	t	920
40mm	t	900

High yield steel bars, diameter

6mm	t	1,200
8mm	t	1,180
10mm	t	1,150
12mm	t	1,120
16mm	t	1,100
20mm	t	950
25mm	t	930
32mm	t	920
40mm	t	900

Steel fabric, weight

3-4g/m2	m2	5.00
4-5g/m2	m2	8.00
5-6g/m2	m2	9.00
6-7g/m2	m2	12.00
7-8g/m2	m2	14.00
8-10g/m2	m2	16.00

£

Joints

Open surface plain with cork
filler, width 0.5-1m,
thickness

10mm	m2	10.00
20mm	m2	15.00
25mm	m2	20.00

Formed surface plain with cork
filler, width 0.5-1m, thickness

10mm	m2	32.00
20mm	m2	34.00
25mm	m2	36.00

CLASS H: PRECAST CONCRETE

Copings, sills and weir blocks,
weathered and throated, size

150 x 75mm	m	26.00
200 x 75mm	m	28.00
300 x 75mm	m	30.00

CLASS I: PIPEWORK – PIPES
CLASS J: PIPEWORK – FITTINGS AND VALVES
CLASS K: PIPEWORK – MANHOLES AND PIPEWORK
ANCILLARIES
CLASS L: PIPEWORK – SUPPORTS AND PROTECTION

Trench excavation

Excavating trenches for pipe
diameter 225mm, backfilling
and removing surplus from site

trench depth

500mm	m	6.00
1000mm	m	15.00
1500mm	m	30.00
2000mm	m	45.00

£

2500mm	m	65.00
3000mm	m	80.00
3500mm	m	95.00
4000mm	m	110.00
4500mm	m	115.00
5000mm	m	120.00

Excavating trenches for pipe
diameter 300mm, backfilling
and removing surplus from site

trench depth

500mm	m	8.00
1000mm	m	18.00
1500mm	m	32.00
2000mm	m	48.00
2500mm	m	70.00
3000mm	m	84.00
3500mm	m	100.00
4000mm	m	115.00
4500mm	m	120.00
5000mm	m	130.00

Excavating trenches for pipe
diameter 400mm, backfilling
and removing surplus from site

trench depth

1500mm	m	50.00
2000mm	m	70.00
2500mm	m	90.00
3000mm	m	120.00
3500mm	m	140.00
4000mm	m	160.00
4500mm	m	170.00
5000mm	m	180.00

£

Excavating trenches for pipe
diameter 525mm, backfilling
and removing surplus from site

trench depth		
1500mm	m	60.00
2000mm	m	80.00
2500mm	m	100.00
3000mm	m	130.00
3500mm	m	150.00
4000mm	m	170.00
4500mm	m	180.00
5000mm	m	190.00

Excavating trenches for pipe
diameter 900mm, backfilling
and removing surplus from site

trench depth		
1500mm	m	100.00
2000mm	m	135.00
2500mm	m	165.00
3000mm	m	200.00
3500mm	m	235.00
4000mm	m	270.00
4500mm	m	300.00
5000mm	m	340.00

Beds, haunches and surrounds

Bed of sand 150mm thick to pipe
diameter

225mm	m	7.00
300mm	m	8.00
400mm	m	10.00
525mm	m	15.00
900mm	m	20.00

£

Bed of granular material
150mm thick to pipe
diameter

225mm	m	6.00
300mm	m	7.00
400mm	m	9.00
525mm	m	14.00
900mm	m	18.00

Bed of concrete 150mm
thick to pipe diameter

225mm	m	18.00
300mm	m	20.00
400mm	m	22.00
525mm	m	28.00
900mm	m	38.00

Bed and haunching of
concrete to pipe diameter

225mm	m	40.00
300mm	m	48.00
400mm	m	56.00
525mm	m	72.00
900mm	m	116.00

Bed and surround of sand
to pipe diameter

225mm	m	24.00
300mm	m	28.00
400mm	m	46.00
525mm	m	68.00
900mm	m	146.00

£

Bed and surround of
granular material
to pipe diameter

225mm	m	18.00
300mm	m	24.00
400mm	m	30.00
525mm	m	46.00
900mm	m	78.00

Bed and of concrete
to pipe diameter

225mm	m	54.00
300mm	m	66.00
400mm	m	86.00
525mm	m	120.00
900mm	m	240.00

Pipes laid in prepared trenches

Vitrified clay to BS65, spigot and
socket joints with sealing ring,
nominal bore

225mm	m	38.00
300mm	m	56.00
400mm	m	110.00
450mm	m	140.00

Vitrified clay to BS65, plain
ended with sleeve joints, nominal
bore

225mm	m	44.00
300mm	m	66.00

£

Concrete pipes Class L with
rebated flexible plain ended
joints, nominal bore

225mm	m	20.00
300mm	m	25.00
450mm	m	30.00
525mm	m	36.00
750mm	m	70.00
900mm	m	95.00
1200mm	m	140.00
1500mm	m	280.00
1800mm	m	320.00

Ductile spun iron pipes
with spigot and socket Tyton
joints, nominal bore

100mm	m	32.00
150mm	m	38.00
250mm	m	62.00
400mm	m	120.00
600mm	m	190.00

Ductile spun iron pipes
with Stantyte joints,
normal bore

800mm	m	200.00
900mm	m	220.00
1000mm	m	280.00
1200mm	m	380.00
1600mm	m	700.00

£

Carbon steel pipes with
welded joints, normal bore

100mm	m	30.00
150mm	m	38.00
200mm	m	46.00
250mm	m	52.00
300mm	m	58.00

Unplasticised PVC pipes
with ring seal joints,
nominal bore

82mm	m	14.00
110mm	m	12.00
160mm	m	20.00

Unplasticised perforated
PVC pipes, with ring seal
joints, nominal bore

82mm	m	12.00
110mm	m	10.00
160mm	m	22.00

Manholes

Precast concrete manholes
complete including excavation
concrete surround, base and
cover slab, channels, step
irons and inspection cover

675mm internal diameter

depth, 1m	nr	750.00
depth, 1.5m	nr	850.00
depth, 2m	nr	950.00
depth, 2.5m	nr	1100.00
depth, 3m	nr	1200.00
depth, 3.5m	nr	1300.00
depth, 4m	nr	1400.00

£

900mm internal diameter

depth, 1m	nr	850.00
depth, 1.5m	nr	980.00
depth, 2m	nr	1100.00
depth, 2.5m	nr	1200.00
depth, 3m	nr	1300.00
depth, 3.5m	nr	1400.00
depth, 4m	nr	1500.00
depth, 4.5m	nr	1650.00
depth, 5m	nr	1850.00

1200mm internal diameter

depth, 1.5m	nr	1400.00
depth, 2m	nr	1650.00
depth, 2.5m	nr	1850.00
depth, 3m	nr	2000.00
depth, 3.5m	nr	2150.00
depth, 4m	nr	2300.00
depth, 4.5m	nr	2450.00
depth, 5m	nr	2600.00

1500mm internal diameter

depth, 2m	nr	1800.00
depth, 2.5m	nr	2000.00
depth, 3m	nr	2150.00
depth, 3.5m	nr	2300.00
depth, 4m	nr	2450.00
depth, 4.5m	nr	2600.00
depth, 5m	nr	2750.00

1800mm internal diameter

depth, 2m	nr	2100.00
depth, 2.5m	nr	2500.00
depth, 3m	nr	3000.00
depth, 3.5m	nr	3300.00
depth, 4m	nr	3600.00
depth, 4.5m	nr	4000.00
depth, 5m	nr	4400.00

£

Gullies

Vitrified clay road gully,
480mm diameter x 900mm
deep, including excavation,
concrete surround, engineering
brick seating and cast iron
grating   nr   340.00

Precast concrete road gully,
375mm diameter x 900mm deep,
including excavation, concrete
surround, Class B engineering
brick seating and cast iron grating   nr   250.00

CLASS M: STRUCTURAL METALWORK

Frabrication of members
for frames

columns	t	1000.00
beams	t	1000.00
portal frames	t	900.00
trusses and built up girders	t	1250.00
bracings, purlins and cladding rails	t	1300.00
anchorages and holding down assemblies	t	2250.00

Permanent erection of
members for frames   t   250.00

Site bolts

black	t	2750.00
HSFG general grade	t	2750.00
HSFG higher grade	t	3000.00
HSFG load indicating or load limit	t	4000.00

£

Offsite surface treatment

blast cleaning	m2	5.00
wire brushing	m2	3.00
one coat chromate primer	m2	3.00
galvanizing	m2	15.00

CLASS N: MISCELLANEOUS METALWORK

Stairways and landings	t	4250.00
Walkways and platforms	t	4000.00

Cat ladder in galvanized steel
rungs at 300mm centres,
strings extended to form
handrail, 450mm wide,
length

3m	nr	900.00
4m	nr	1200.00
5m	nr	1500.00
6m	nr	1800.00
7m	nr	2100.00
8m	nr	2400.00
9m	nr	2700.00
10m	nr	3000.00

Guard cage to cat ladder	m	80.00

Galvanised steel staircase
900mm wide with chequer
plate treads balustrade one
side, supported on universal
columns

5500mm going, 3000mm rise, 16 treads and one landing	nr	3000.00
10000mm going, 5000mm rise, 16 treads and one landings	nr	3500.00

		£
Galvanised tubular handrail, 1050mm high, standards at 2000mm centres, with middle rail	m	125.00
Galvanised flat section handrail and members, standards at 1000mm centres, infilled with square vertical bars at 100mm centres	m	150.00

Safety fencing

Tensioned corrugated to DTp Clause 409 with "Z" section steel posts set in concrete

single sided	m	30.00
double sided	m	45.00

Untensioned corrugated to DTp Clause 412, single sided with timber posts

	m	25.00

Miscellaneous framing

Angle section

150 x 75 x 10mm	m	25.00
100 x 100 x 10mm	m	28.00
150 x 150 x 12mm	m	32.00

Channel section

150 x 75 x 10mm	m	35.00
250 x 75 x 16mm	m	40.00

Flooring

Galvanized mild steel 'Durbar' pattern plate 8mm thick	m2	120.00
Galvanized open grid flooring 50mm thick	m2	60.00

£

CLASS O: TIMBER

Greenheart timber in
marine works

100 x 75mm	m	22.00
150 x 75mm	m	24.00
200 x 200mm	m	60.00
200 x 300mm	m	80.00
300 x 300mm	m	110.00
600 x 600mm	m	280.00

Wrought softwood in
marine work

100 x 75mm	m	15.00
150 x 75mm	m	18.00
200 x 200mm	m	40.00
200 x 300mm	m	65.00
300 x 300mm	m	75.00
600 x 600mm	m	200.00

Hardwood decking,
thickness

50mm	m	100.00
75mm	m	130.00
100mm	m	175.00

Softwood decking,
thickness

50mm	m	50.00
75mm	m	70.00
100mm	m	100.00

£

Metalwork

Coach screws, 10mm
Diameter, length

75mm	nr	2.00
100mm	nr	3.00
150mm	nr	4.00

Blackbolts, nuts and
washers, M12, length

100mm	nr	3.00
140mm	nr	4.00
200mm	nr	5.00

CLASS P: PILING
CLASS Q: PILING ANCILLARIES

Bored cast-in-place concrete piles

Allow £6,000 and £10,000 for the cost of setting up and removing
from site for 50 piles and 100 piles respectively.

Reinforced in situ concrete
piles 300mm diameter

concrete	m	20.00
depth bored, 10m	m	45.00
depth bored, 15m	m	50.00
depth bored, 20m	m	55.00
depth bored, 25m	m	60.00

Reinforced in situ concrete
piles 450mm diameter

concrete	m	35.00
depth bored, 10m	m	90.00
depth bored, 15m	m	100.00
depth bored, 20m	m	110.00
depth bored, 25m	m	120.00

£

Reinforced in situ concrete
piles 600mm diameter

concrete	m	50.00
depth bored, 10m	m	170.00
depth bored, 15m	m	180.00
depth bored, 20m	m	190.00
depth bored, 25m	m	210.00

Driven cast-in-place concrete piles

Allow £3,500 and £7,000 for the cost of setting up and removing from site for 50 piles and 100 piles respectively.

Driven concrete piles
300mm diameter

concrete	m	20.00
depth driven, 10m	m	70.00
depth driven, 15m	m	110.00
depth driven, 20m	m	140.00
depth driven, 25m	m	175.00

Driven concrete piles
450mm diameter

concrete	m	35.00
depth driven, 10m	m	90.00
depth driven, 15m	m	135.00
depth driven, 20m	m	180.00
depth driven, 25m	m	225.00

Driven concrete piles
600mm diameter

concrete	m	50.00
depth driven, 10m	m	150.00
depth driven, 15m	m	225.00
depth driven, 20m	m	300.00
depth driven, 25m	m	375.00

£

Steel sheet piling

Frodingham steel piles, ref.1N, driven into ground	m2	90.00
extra for corner sections	m	70.00
extra for junctions	m	95.00
Frodingham steel piles, ref.2N, driven into ground	m2	100.00
extra for corner sections	m	70.00
extra for junctions	m	95.00
Frodingham steel piles, ref.3N, driven into ground	m2	110.00
extra for corner sections	m	70.00
extra for junctions	m	95.00
Frodingham steel piles, ref.4N, driven into ground	m2	140.00
extra for corner sections	m	70.00
extra for junctions	m	95.00

CLASS R: ROADS AND PAVINGS

Sub-bases flexible road bases and Surfacing

Hardcore road base	m3	22.00
depth, 100mm	m2	3.00
depth, 150mm	m2	4.00
depth, 200mm	m2	4.00
depth, 250mm	m2	6.00

£

Granular material, DTp type 1	m3	26.00
depth, 100mm	m2	4.00
depth, 150mm	m2	5.00
depth, 200mm	m2	6.00
depth, 250mm	m2	7.00
Granular material, DTp type 2	m3	25.00
depth, 100mm	m2	4.00
depth, 150mm	m2	5.00
depth, 200mm	m2	6.00
depth, 250mm	m2	7.00

Dense bitumen macadam, DTp
clause 908, 14mm aggregate,
base course, depth

30mm	m2	5.00
70mm	m2	9.00

Dense bitumen macadam, DTp
clause 913, wearing course, depth

30mm	m2	5.00
50mm	m2	8.00

Concrete pavements

Carriageway slabs, concrete
grade C20, depth

150mm	m2	15.50
200mm	m2	20.40
250mm	m2	25.50
300mm	m2	30.60

Steel fabric reinforcement
to BS4483

reference A142 2.22 kgs/m2	m2	5.50
reference B503 5.93 kgs/m2	m2	9.50

£

Waterproof membranes
below concrete pavements,
plastic sheeting, 1200 gauge   m2   2.80

Joints in concrete pavements

Longitudinal joints, 10mm
diameter x 750mm long mild steel
dowels at 750mm centres, sealed
with polysulphide, depth

150mm	m	26.00
220mm	m	32.00
250mm	m	36.00

Kerbs, channels and edgings

Precast concrete kerbs to BS340
straight or curved to radius
exceeding 12m

150 x 305mm	m	15.00
125 x 255mm	m	12.00

Concrete in bed to kerbs

200 x 150mm	m	4.20
300 x 150mm	m	6.00
400 x 150mm	m	8.00
150 x 150mm haunching	m	1.50

Drop kerbs

125 x 255mm	nr	12.00
150 x 305mm	nr	8.00

Precast concrete channels
straight or curved to radius
exceeding 12m

255 x 125mm	m	12.00

£

Precast concrete edging
straight or curved to radius
exceeding 12m

150 x 50mm	m	6.00

Road markings

Reflectorised white

continuous lines

150mm wide	m	1.00
200mm wide	m	1.20

broken lines

100mm wide, 1m line 3m gap	m	1.00
100mm wide, 2m line 5m gap	m	1.10
100mm wide, 4m line 2m gap	m	1.20

arrow

4m length, straight	nr	22.00
4m length, turned	nr	24.00
6m length, straight	nr	28.00
6m length, turned	nr	30.00

letters

1.6m high	nr	10.00
2.0m high	nr	18.00
3.0m high	nr	20.00

140 x 250mm reflecting studs with cats' eye reflection	nr	14.00

CLASS S: RAIL TRACK

Track foundation

Bottom ballast granite, crushed graded 50-25mm	m3	50.00
Top ballast granite, crushed graded 50-25mm	m3	60.00

£

Taking up track and turnouts, dismantle and stack

Bullhead or flat bottom rail

plain track, timber sleepers, fishplate joints	m	8.00
turnouts, timber sleepers, fishplate joints	nr	400.00
plain track, concrete sleepers, fishplate joints	m	10.00
turnouts, concrete sleepers, fishplate joints	nr	450.00

Lifting packing and slewing

Bullhead rail track on timber
sleepers track length 10m,
maximum slew 300mm,
maximum lift 100mm   nr   550.00

Supplying only plain line material

Bullhead rails; for joints or welded track

mass 40-50 kg/m, section reference 95R	t	900.00

Sleepers, softwood timber, 250 x 125 x 2600mm long	nr	50.00
Sleepers, hardwood timber, 250 x 125 x 2600mm long	nr	60.00

Sleepers, concrete type 'F27' with 2 nr cast iron 'Pandrol' fittings cast in	nr	35.00

Fittings

Chairs, cast iron 'CC' pattern	nr	60.00
Fish plates, standard set	nr	60.00

		£
Fish plates, insulated set	nr	60.00
Switches and crossings		
turnouts	nr	18.00
diamond crossings	nr	90.00
Buffer stops, 2-2.5 tonnes	nr	2500.00

Laying only plain line material

Bullhead rails

plain track, mass 40-50kg/m fish plate joints, timber sleepers	m	45.00
welded joints, concrete sleepers	m	50.00
turnout, standard type on timber sleepers	nr	2250.00
buffer stop, single rake	nr	250.00

CLASS T: TUNNELS

There are many different methods of
boring and constructing tunnels so
the following rates should treated
with caution. For costs on specific
jobs the advice of a tunnelling contractor
should be sought.

Excavation

Tunnels in rock, straight

diameter 1.8m	m3	450.00
diameter 3.0m	m3	300.00

£

Tunnels in clay, straight

diameter 1.8m	m3	250.00
diameter 3.0m	m3	150.00

Shafts in rock, vertical

diameter 3.0m	m3	200.00
diameter 4.5m	m3	150.00

Shafts in clay, vertical

diameter 3.0m	m3	120.00
diameter 4.5m	m3	100.00

In situ lining to tunnels, vertical

Cast concrete grade C20, diameter 2m	m3	250.00

In situ lining to shafts, vertical

Cast concrete grade C20, diameter 2m	m3	225.00

Formwork, rough finish
diameter 2m   m2   80.00

Formwork, smooth finish
diameter 2m   m2   85.00

**Pre-formed segmental linings
to tunnels**

Precast concrete bolted
flanged rings, depth 450mm

diameter 2.5m	nr	500.00
diameter 3.m	nr	900.00

£

**Pre-formed segmental linings
to shafts**

Precast concrete bolted
flanged rings, depth 450mm

diameter 3m	nr	800.00
diameter 4.5mm	nr	1100.00

CLASS U: BRICKWORK, BLOCKWORK AND MASONRY

Common brickwork

Brickwork (£150 per 1000) in
cement mortar (1:3)

vertical walls		
102.5mm thick	m2	55.00
215mm thick	m2	90.00
327mm thick	m2	120.00
440mm thick	m2	150.00

In columns and piers

215m x 215mm	m	30.00
327 x 327mm	m	35.00
440 x 440mm	m	40.00

Facing brickwork

Brickwork (£300 per 1000) in
gauged mortar (1:1:6)

vertical walls		
102.5mm thick	m2	45.00
215mm thick	m2	85.00

£

Engineering brickwork
(£200 per 1000) in cement
mortar (1:3)

vertical walls			
215mm thick		m2	75.00
327mm thick		m2	110.00
440mm thick		m2	175.00

Engineering brickwork
PC £300 per 1000) in cement
mortar (1:3)

facing to vertical walls			
102.5mm thick		m2	55.00
215mm thick		m2	85.00

Lightweight blockwork

Blockwork in cement
mortar (1:3) in vertical
straight walls

100mm thick	m2	24.00
140mm thick	m2	30.00
190mm thick	m2	36.00

Dense concrete blockwork

Blockwork in cement
mortar (1:L3) in vertical
straight walls

100mm thick	m2	30.00
140mm thick	m2	40.00
190mm thick	m2	48.00

£

Ashlar masonry

Portland Whitbed with one
exposed face in gauged mortar
(1:1:6)

facing to vertical walls		
50mm thick	m2	220.00
75mm thick	m2	300.00

CLASS V: PAINTING
(Rates inclusive of all inclinations)

One coat primer on general
surfaces exceeding 300mm

metal	m2	3.00
timber	m2	3.00

Two coats emulsion paint
on general surfaces
exceeding 300mm

smooth concrete	m2	7.00
blockwork and brickwork	m2	8.00

Two coats of cement painting

smooth concrete	m2	7.00
blockwork and brickwork	m2	8.00
rough cast surfaces	m2	9.00

Three coats of oil paint on

primed steel sections	m2	9.00
primed pipe work	m2	10.00
primed timber	m2	11.00

£

CLASS W: WATERPROOFING

Damp proofing

One layer 1000 gauge 'Bituthene' sheet, fixed with adhesive	m2	8.00
Asphalt to BS1097 two coat work, 20mm thick on concrete surfaces	m2	30.00

Roofing

Asphalt to BS908, two coat work, 20mm thick on concrete surfaces	m2	32.00
Built up felt roofing to BS747, three layer coverings	m2	25.00
Protective layers, one layer 1000 gauge polythene sheet, fixed with adhesive	m2	3.00
Cement and sand (1:3) screed with waterproof additive	m2	15.00
Sprayed or brushed waterproofing two coats of 'Synthaprufe' to concrete sufaces	m2	6.00

CLASS X: MISCELLANEOUS WORK

See Landscaping section

£

CLASS Y: SEWER RENOVATION AND ANCILLARY WORKS

Preparation of existing sewer

Cleaning egg shaped sewer 1050mm high	m	15.00

Removing intrusions

brickwork	m3	75.00
concrete	m3	120.00

Plugging laterals with concrete

bore not exceeding 300mm	nr	75.00
bore exceeding 300mm	nr	125.00

Local internal repairs

area 0.1.25m2	nr	50.00
area 5m2	nr	200.00

Stabilisation of existing sewers

Pointing with cement mortar (1:3)	m2	30.00

Renovation of existing sewers

Segmental lining in GRP

egg shaped 1050mm high	m	320.00

Laterals to renovated sewers

Jointing

bore not exceeding 150mm	nr	60.00
bore 150-300mm	nr	110.00
bore 450mm	nr	170.00

£

Interruptions

Preparation of existing sewers

cleaning	hour	325.00

Stabilisation

pointing	hour	75.00

Renovation

linings	hour	80.00

CLASS Z: SIMPLE BUILDING WORKS INCIDENTAL TO CIVIL ENGINEERING WORKS

See Building work

MECHANICAL WORK

£

Pipework

Mild steel pipes, heavy, black,
malleable iron, fixing with
standard supports, including
fittings

screwed fittings, pipe size			
20mm		m	30.00
25mm		m	34.00
32mm		m	38.00
40mm		m	42.00
50mm		m	50.00
65mm		m	60.00
80mm		m	72.00
110mm		m	98.00
125mm		m	120.00
150mm		m	148.00
welded fittings, pipe size			
20mm		m	28.00
25mm		m	32.00
32mm		m	36.00
40mm		m	40.00
50mm		m	46.00
65mm		m	54.00
80mm		m	64.00
110mm		m	90.00
125mm		m	110.00
150mm		m	138.00

Mild steel pipes, medium, black,
malleable iron, fixing with
standard supports, including
fittings

screwed fittings, pipe size			
20mm		m	28.00
25mm		m	32.00
32mm		m	36.00

		£
40mm	m	40.00
50mm	m	48.00
65mm	m	58.00
80mm	m	70.00
110mm	m	96.00
125mm	m	118.00
150mm	m	146.00

welded fittings, pipe size

20mm	m	26.00
25mm	m	30.00
32mm	m	34.00
40mm	m	38.00
50mm	m	46.00
65mm	m	52.00
80mm	m	62.00
110mm	m	88.00
125mm	m	108.00
150mm	m	136.00

Copper pipes: B2871 part 1
table X, fixing with standard
supports, including fittings

capillary fittings, pipe size

15mm	m	14.00
22mm	m	16.00
28mm	m	20.00
35mm	m	32.00
42mm	m	40.00
54mm	m	48.00

compression fittings, pipe size

15mm	m	18.00
22mm	m	20.00
28mm	m	24.00
35mm	m	38.00
42mm	m	48.00
54mm	m	56.00

£

Carbon steel pipes to BS3601,
fixing with standard supports,
including fittings

welded fittings, pipe size		
200mm	m	160.00
250mm	m	210.00
300mm	m	240.00
350mm	m	310.00
400mm	m	520.00

Ductwork

Rectangular galvanised sheet
steel ductwork, supports and doors

girth of 2 sides		
250mm	m	70.00
500mm	m	75.00
750mm	m	80.00
1000mm	m	90.00
1250mm	m	95.00
1500mm	m	110.00
1750mm	m	130.00
2000mm	m	160.00
2250mm	m	170.00
2500mm	m	210.00
2750mm	m	240.00
3000mm	m	300.00

Circular galvanised steel
Spirally wound ductwork
Supports and doors, diameter

100mm	m	32.00
160mm	m	34.00
200mm	m	45.00
250mm	m	48.00
315mm	m	52.00
355mm	m	55.00
400mm	m	60.00
450mm	m	65.00

		£
500mm	m	70.00
630mm	m	80.00
710mm	m	85.00
800mm	m	95.00
900mm	m	110.00
1000mm	m	130.00

Thermal insulation

Rigid mineral wool sections
bright class 'O' foil covered,
secured with aluminium bands at
300mm centres, including for
fixing around joints, flanges,
valves and the like.

Insulation thickness 25mm,
pipe size

15mm	m	8.00
20mm	m	8.50
25mm	m	9.00
32mm	m	9.50
40mm	m	10.00
50mm	m	11.00
65mm	m	12.00
80mm	m	14.00
100mm	m	18.00

Insulation thickness 40mm,
pipe size

15mm	m	10.00
20mm	m	10.50
25mm	m	11.00
32mm	m	11.50
40mm	m	12.00
50mm	m	13.00
65mm	m	14.00
80mm	m	16.00
100mm	m	20.00

£

Insulation thickness 50mm,
pipe size

15mm	m	12.00
20mm	m	12.50
25mm	m	13.00
32mm	m	13.50
40mm	m	14.00
50mm	m	15.00
65mm	m	16.00
80mm	m	18.00
100mm	m	22.00

Boilers

Domestic gas fired central
heating boilers, floor mounted,
balanced flue

9-12kW	nr	850.00
12-15kW	nr	880.00
15-18kW	nr	920.00
18-21kW	nr	950.00
21-23kW	nr	1050.00
23-29kW	nr	1350.00
29-37kW	nr	1500.00
37-41kW	nr	1550.00

Domestic gas fired central
heating boilers, wall mounted,
balanced flue

6-9kW	nr	750.00
19-12kW	nr	820.00
12-15kW	nr	930.00
15-18kW	nr	1050.00
18-22kW	nr	1100.00

Commercial packaged sectional
floor mounted gas fired boiler,
pressure jet burner connected
to conventional flue complete

16-26kW	nr	2300.00
23-33kW	nr	2400.00

£

33-40kW	nr	2700.00
35-50kW	nr	3100.00
50-65kW	nr	3400.00
65-80kW	nr	4000.00
80-100kW	nr	5000.00
100-120kW	nr	5750.00
105-140kW	nr	6500.00
140-180kW	nr	7000.00
180-230kW	nr	7300.00
230-280kW	nr	7750.00
280-330kW	nr	8500.00

Commercial packaged floor mounted oil fired hot water boiler, complete

130-190kW	nr	7000.00
200-250kW	nr	7250.00
280-360kW	nr	8500.00
375-500kW	nr	9500.00
580-730kW	nr	12250.00
655-820kW	nr	12400.00
830-1040kW	nr	13000.00
1070-1400kW	nr	16400.00
1420-1850kW	nr	21000.00
1850-2350kW	nr	24000.00
2300-3000kW	nr	28000.00
2800-3500kW	nr	36000.00

Radiators

Pressed steel single panel radiator 450mm high complete with valves, length

500mm	m	170.00
600mm	m	178.00
800mm	m	186.00
1000mm	m	190.00
1200mm	m	200.00
1400mm	m	210.00
1600mm	m	215.00
1800mm	m	250.00
2000mm	m	280.00

£

Pressed steel single panel
radiator 700mm high complete
with valves, length

500mm	m	180.00
600mm	m	200.00
800mm	m	220.00
1000mm	m	240.00
1200mm	m	260.00
1600mm	m	280.00
1800mm	m	300.00
2000mm	m	320.00

Pressed steel double panel
radiator 450mm high complete
with valves, length

500mm	m	180.00
600mm	m	200.00
800mm	m	210.00
1000mm	m	220.00
1200mm	m	250.00
1600mm	m	280.00
1800mm	m	330.00
2000mm	m	380.00

Pressed steel double panel
radiator 700mm high complete
with valves, length

500mm	m	210.00
600mm	m	240.00
800mm	m	280.00
1000mm	m	300.00
1200mm	m	320.00
1400mm	m	370.00
1600mm	m	420.00
1800mm	m	460.00
2000mm	m	500.00

ELECTRICAL WORK

£

Transformers

Transformer, 11kV/415 volt,
50 Hz three phase, air cooled,
oil filled, skid mounted, cable
boxes, fixing to backgrounds

500kVA	nr	7500.00
800kVA	nr	8500.00
1000kVA	nr	9500.00
1250kVA	nr	11500.00
1500kVA	nr	13500.00
2000kVA	nr	18000.00

Transformer, 11kV/415 volt,
50 Hz three phase, hermetically,
sealed, silicone, impregnated,
skid mounted, cable boxes,
fixing to backgrounds

500kVA	nr	10000.00
800kVA	nr	11250.00
1000kVA	nr	13000.00
1250kVA	nr	15250.00
1500kVA	nr	18000.00
2000kVA	nr	20000.00

Extra for

fluid temperature indicator	nr	350.00
plain roller	nr	250.00
pressure relief device	nr	450.00

£

Distribution boards

MCB distribution board, steel casting, 125/25amp
incorners, fixed to backgrounds

SP&N			
	6 way	nr	120.00
	8 way	nr	140.00
	12 way	nr	160.00
	16 way	nr	180.00
TP&N			
	4 way	nr	500.00
	6 way	nr	520.00
	8 way	nr	560.00
	12 way	nr	580.00

Circuit breakers

Residual current circuit breakers for distribution
boards including connection

SP&N			
	10mA 6Amp	nr	80.00
	10mA 10-32Amp	nr	80.00
	10mA 6Amp 45Amp	nr	80.00
	30mA 6Amp	nr	85.00
	30mA 10-40Amp	nr	85.00
	30mA 50-63Amp	nr	85.00
	100mA 6Amp	nr	135.00
	100mA 10-40Amp	nr	135.00
	100mA 50-63Amp	nr	135.00

Busbar trunking

Rising mains busbar, insulated supports,
earth continuity bar, fixed to backgrounds

200 Amp TP&N		m	200.00
	end cap	nr	30.00
	end feed unit	nr	350.00
	top feed unit	nr	350.00
	flat tee	nr	220.00

£

315 Amp TP&N	m	220.00
end cap	nr	33.00
end feed unit	nr	360.00
top feed unit	nr	360.00
flat tee	nr	240.00
400 Amp TP&N	m	240.00
end cap	nr	40.00
tee flat edge	nr	400.00
top feed unit	nr	400.00
flat tee	nr	340.00
630 Amp TP&N	m	300.00
end cap	nr	45.00
end feed unit	nr	420.00
top feed unit	nr	420.00
flat tee	nr	350.00
800 Amp TP&N	m	410.00
end cap	nr	95.00
end feed unit	nr	480.00
top feed unit	nr	480.00
flat tee	nr	490.00

HV/LV cables and wiring

PVC sheathed cable,
6350/11000 volt grade, lead-
covered, paper-insulated single
wire armoured, with copper
conductors

three core cable, clipped to backgrounds		
95mm2	m	34.00
120mm2	m	38.00
150mm2	m	44.00
185mm2	m	48.00
240mm2	m	52.00
300mm2	m	62.00

£

Terminations

三 core cable, clipped
to backgrounds

95mm2	nr	600.00
120mm2	nr	630.00
150mm2	nr	670.00
185mm2	nr	680.00
240mm2	nr	720.00
300mm2	nr	780.00

LSF sheathed light duty
XLPE insulated cable
600 volt grade with copper
conductors, fixed to backgrounds

two core

2L 1.5mm2	m	3.00
2L 2.5mm2	m	3.50
3L 4.0mm2	m	4.50
3L 6.0mm2	m	5.00
4L 10.0mm2	m	6.50
4L 16.0mm2	m	8.50
7L 25.0mm2	m	9.00
7L 35.0mm2	m	10.00

three core

2L 1.5mm2	m	4.00
2L 2.5mm2	m	4.50
3L 4.0mm2	m	5.50
3L 6.0mm2	m	6.00
4L 10.0mm2	m	7.50
4L 16.0mm2	m	9.00
7L 25.0mm2	m	10.00
7L 35.0mm2	m	11.00

four core

2L 2.5mm2	m	5.50
3L 4.0mm2	m	6.50
4L 10.0mm2	m	8.50
4L 16.0mm2	m	10.00
7L 35.0mm2	m	12.00

£

Cable trays

Galvanised steel, heavy duty
perforated cable trays including
supports, fixings, bends, tees and
reducers

50mm wide	m	9.00
100mm wide	m	13.00
150mm wide	m	14.00
300mm wide	m	20.00
600mm wide	m	34.00
750mm wide	m	45.00
900mm wide	m	55.00

Conduits

Steel conduit including all fittings
and supports

galvanised

20mm	m	9.00
25mm	m	10.00
32mm	m	11.00
38mm	m	14.00
50mm	m	19.00

black enamelled

20mm	m	8.00
25mm	m	9.00
32mm	m	10.00
38mm	m	13.00
50mm	m	18.00

PVC conduit including all fittings
and supports

white light gauge, super
high impact

16mm diameter	m	2.50
20mm diameter	m	3.00
25mm diameter	m	3.50
32mm diameter	m	4.00

£

white light gauge, super
high impact

16mm diameter	m	3.00
20mm diameter	m	3.50
25mm diameter	m	4.00
32mm diameter	m	4.50
38mm diameter	m	5.50
50mm diameter	m	7.50

Trunkings

Galvanised steel trunking
complete with fittings, fixing
to backgrounds

single compartment

50 x 50mm	m	24.00
75 x 50mm	m	26.00
75 x 75mm	m	28.00
100 x 50mm	m	28.00
100 x 75mm	m	30.00
100 x 100mm	m	30.00
150 x 50mm	m	30.00
150 x 75mm	m	32.00
150 x 100mm	m	36.00
150 x 150mm	m	38.00
225 x 150mm	m	44.00
225 x 225mm	m	52.00
300 x 225mm	m	60.00
300 x 300mm	m	68.00

twin compartment

50 x 50mm	m	26.00
75 x 50mm	m	28.00
75 x 75mm	m	30.00
100 x 50mm	m	32.00
100 x 75mm	m	34.00
100 x 100mm	m	36.00
150 x 50mm	m	36.00
150 x 75mm	m	38.00
150 x 100mm	m	42.00
150 x 150mm	m	48.00

£

triple compartment

50 x 50mm	m	30.00
75 x 50mm	m	32.00
75 x 75mm	m	34.00
100 x 50mm	m	36.00
100 x 75mm	m	38.00
100 x 100mm	m	38.00
150 x 50mm	m	46.00
150 x 75mm	m	48.00
150 x 100mm	m	52.00

Luminaires

Surface mounted fluorescent
luminaire, single tube, switch start,
fixing to backgrounds

1253 x 184mm 36watt	nr	80.00
1553 x 184mm 58watt	nr	90.00
1817 x 184mm 70watt	nr	105.00

Surface mounted fluorescent
luminaire, twin tube, switch start,
fixing to backgrounds

1253 x 184mm 36watt	nr	100.00
1553 x 184mm 58watt	nr	115.00
1817 x 184mm 70watt	nr	130.00

Road lighting

High pressure sodium lantern,
aluminium canopy, support,
clear bowl, lighting column,
aluminium outreach bracket,
fixing to base

250 watt 10 metre high	nr	1200.00
400 watt 10 metre high	nr	1300.00
250 watt 12 metre high	nr	1500.00
400 watt 12 metre high	nr	1600.00

£

Escalators

30 degrees incline, 3-4m rise, glass balustrades and 600 mm steps	nr	104000.00
extra for 5-6m rise	nr	15000.00
extra for stainless steel balustrades and decking	nr	5000.00
extra for under handrail lighting	nr	4000.00
30 degrees incline, 4-5 rise, glass balustrades and 800 mm steps	nr	120000.00
extra for 5-6m rise	nr	15000.00
extra for stainless steel balustrades and decking	nr	8000.00
extra for under handrail lighting	nr	5000.00

ALTERATIONS AND REPAIRS

£

Taking down

Take down walls and remove
rubble from site

		£
external walls		
half brick wall	m2	18.00
one brick wall	m2	26.00
one brick and a half wall	m2	38.00
two brick wall	m2	50.00
blockwork 140mm wall	m2	16.00
blockwork 190mm wall	m2	18.00
blockwork 215mm wall	m2	24.00
internal walls		
half brick wall	m2	20.00
one brick wall	m2	28.00
blockwork 90mm wall	m2	10.00
blockwork 100mm wall	m2	11.00
blockwork 115mm wall	m2	13.00
blockwork 125mm wall	m2	15.00
blockwork 140mm wall	m2	18.00

Take down and remove chimney
stack to below roof slop, size

750 x 750 x 1200mm above roof	nr	260.00
1030 x 1030 x 1000mm above roof	nr	380.00

Take down chimney breast from
roof to ground level (2 storey) — nr — 1800.00

£

Alterations

Form openings in walls

in gauged mortar		
half brick wall	m2	60.00
one brick wall	m2	110.00
brick/block cavity wall	m2	108.00
100mm block wall	m2	48.00
215mm block wall	m2	72.00
in cement mortar		
half brick wall	m2	90.00
one brick wall	m2	140.00
brick/block cavity wall	m2	148.00
100mm block wall	m2	58.00
215mm block wall	m2	80.00

Cut opening in internal wall
900 x 2100mm suitable for new single
door including squaring up jambs and
head, inserting lintel over, making
good to surrounding plasterwork

half brick wall	nr	360.00
one brick wall	nr	400.00
one and half brick wall	nr	460.00
100mm block wall	nr	300.00
215mm block wall	nr	360.00

Cut opening in internal wall
1200 x 1200mm for new window
including squaring up jambs,
head and cill, inserting lintel over, making
good surrounding plasterwork

one brick wall	nr	550.00
brick/block cavity wall	nr	620.00

£

Fill openings in common
brickwork in gauged mortar,
including plastering one side

one brick wall	nr	280.00
brick/block cavity wall	nr	380.00

Fill openings in blockwork
in gauged mortar, including
plastering one side

100mm thick blockwork wall	m2	180.00
215mm thick blockwork wall	m2	240.00

Fill openings in concrete floor with
concrete

100mm thick	m2	14.00
150mm thick	m2	21.00
200mm thick	m2	28.00
250mm thick	m2	35.00

Fill openings in suspended concrete
floor with concrete including formwork
to soffit

100mm thick	m2	54.00
150mm thick	m2	60.00
200mm thick	m2	70.00
250mm thick	m2	75.00

Cut out defective facing bricks
in one brick wall and renew

£350 per thousand		
single brick	nr	6.00
areas less than 1m2	m2	122.00
areas over 1m2	m2	116.00

£

£500 per thousand
single brick	nr	8.00
areas less than 1m2	m2	140.00
areas over 1m2	m2	130.00

Cut out raking crack in brickwork,
stitch in new facing brickwork
(PC £350 per thousand) pointing
one side to match existing

half brick thick wall	m	56.00
one brick thick wall	m	88.00

Renew concrete lintel 1500mm
long including one course of
brickwork, inserting new concrete
lintel and making good   nr   140.00

Rake out joints in brickwork and
repoint in gauged mortar   m2   12.00

Cut out decayed structural softwood
timber member and renew

floors
50 x 150mm	m	10.00
50 x 175mm	m	12.00
50 x 200mm	m	14.00

roofs
50 x 150mm	m	12.00
50 x 175mm	m	14.00
50 x 200mm	m	16.00

Remove damaged tile/slate and
replace to match existing

Welsh slate	nr	20.00
concrete interlocking tile	nr	10.00
clay plain tile	nr	9.00

£

Remove damaged tile/slate and
replace with new to match existing

Welsh slate	nr	90.00
concrete interlocking tile	nr	38.00
clay plain tile	nr	34.00

Cut out defective bituminous
felt roofing and renew

areas less than 1m2	m2	40.00
areas over 1m2	m2	30.00

Take up existing flooring and renew

chipboard flooring	m2	18.00
tongued and grooved softwood	m2	38.00

Take up softwood skirting and renew	m	7.00
Ease adjust and re-hang single door	nr	10.00
Ease opening casement or sash window	nr	8.00
Cut out decayed transome or sash window and piece in new section	nr	20.00
Repair sash window, ease, re-cord, replace weights, beads and ironmongery	nr	60.00
Take out timber staircase (single storey height) and fill in opening in timber floor	nr	450.00
Renew tread or riser to timber staircase	nr	28.00
Take up defective stair nosing and replace with aluminium nosing	nr	22.00

		£
Clean out rainwater gutters	m	2.00
Replace galvanised metal support bracket to gutter	nr	8.00
Take down damaged section of rainwater gutter and replace with new half round PVC-U gutter	m	20.00
Hack off addled plaster and renew		
areas less than 1m2	m2	22.00
areas over 1m2	m2	20.00
Cut out crack to plastered wall and make good	m	6.00
Cut out crack to plastered ceiling and make good	m	6.00
Repair defective floor screed		
areas less than 1m2	m2	18.00
areas over 1m2	m2	16.00
Prepare and refix loose ceramic wall tiles	m2	22.00
Strip off wall coverings, rub down plastered walls and prepare for new finish	m2	3.00
Clean off existing Artex from ceiling and prepare to receive new finish	m2	7.00
Clean down painted timber frames and prepare to receive new finish	m2	2.00
Burn off paint from timber frames, rub down and prepare for new finish	m2	5.00

5

Indices and regional variations

Indices

Indices have been used for many years ion the construction industry as a tool in the comparison of costs and tender prices between different periods of time. They are essential when projecting cash flow forecasts on projects that may over run or for planning medium- to long-term projects where trends from historical cost data bases can be identified and projected forwards.

There are several indices available and the most appropriate must be selected to suit the project under consideration. The main set of indices in the building industry is produced by the Building Cost Information Services (BCIS) and is published by the RICS. The civil engineering industry sometimes use a price adjustment formula (usually called the Baxter formula after its originator) on contracts with a long construction period. The formula is based on fewer headings than the BCIS and covers eleven categories.

Construction indices usually present data under two headings; construction costs and tender prices. Construction costs represent the amounts paid by contractors for labour, materials, plant and other costs incurred in running a business. Tender prices are the sums accepted by clients for the erection of projects and include profit and overheads.

Tender prices tend to increase when there is a large volume of work available and decrease during lean periods. In difficult times, contractors may submit tenders at cost (or even below cost) in order to obtain work to keep the work force together and keep the plant occupied.

Tender prices are compiled from accepted tenders based on statistics produced in quarterly periods and consolidated into annual figures. The following table lists the construction costs and tender prices between 1976 and 2006 – the figures from 2003 and later are estimated.

Year	Construction costs	Tender prices
1976	100	100
1977	114	107
1978	123	124
1979	142	154
1980	170	190
1981	190	194
1982	210	192

Year	Construction costs	Tender prices
1984	236	210
1985	248	218
1986	262	229
1987	276	258
1988	293	309
1989	314	340
1990	337	309
1991	355	262
1992	365	243
1993	372	235
1994	382	252
1995	401	265
1996	411	267
1997	421	283
1998	439	313
1999	456	332
2000	474	348
2001	491	372
2002	520	400
2003	550*	416*
2004	575*	428*
2005	606*	444*
2006	639*	464*

*Estimated

Regional variations

Construction costs vary in indifferent parts of UK and the following adjustments should be made to first stage estimates. The costs in this book are set at 100.

England	East Anglia	83
	East Midlands	80
	Inner London	104
	North	81
	North West	84
	Outer London	98
	South East	92
	South West	86
	Yorkshire and Humberside	81
Northern Ireland		65
Scotland		82
Wales		80

6

Property insurance

When preparing first stage estimates, it is important that the fullest possible financial picture is presented to the client to enable him to make a decision on whether to proceed with the project or not.

A calculation must be done for insurance purposes to assess the rebuilding costs if the building was damaged by fire or some other cause.

The easiest way to achieve this is by using the square metre prices in Chapter 1 as a base. The appropriate rate should be multiplied by the area of the building to be insured. The resultant figure must then be adjusted by both the indices and regional variation factors in Chapter 5.

The insurance cover must also include for the demolition of the damaged building (not just clearing away debris but grubbing up existing foundations and basements) and professional fees to plan and supervise the work of reconstruction. Here is an example of how the calculation should be made.

Example

Health centre in the East Midlands region constructed in 1975 where the original cost is unknown.

	£
Present day cost (Chapter 1) 600 m2 @ say £1,450	870,000
Demolition of old building including site clearance (Chapter 4)	<u>16,800</u>
	<u>886,800</u>
Regional variation (Chapter 5) for Yorkshire and Humberside × 80% × £886,800	709,440
Inflation during the demolition and rebuilding period based on a 18 month construction period and assuming an initial 12 month pre-building period for the preparation of contract documents	
say 4% x 2.5 years	<u>70,944</u>

	Brought forward £		780,384

Fees (Chapter 7)

Professional fees say 12%	93,646		
Planning and building			
Regulations say	25,000		118,646

		£	899,030
	VAT @ 17.5%		157,330
		£	1,056,360

The building should be insured for £1,060,000

7

Professional fees

The use of fees scales used to be mandatory but now clients are able to negotiate the level of fees with their professional advisers.

In the preparation of first stage estimates, the cost of fees must be included on either an all-in fee basis for the professional team or separate fees for each discipline. This chapter contains information for either method.

An allowance should be made for expense incurred in connection with the contract and VAT on professional fees should also be included. The following categories of fees are included:

- Architects
- Quantity Surveyors
- Consulting Engineers
- Landscape Architects
- Planning
- Building Regulations.

Architects' fees

Details of Architects' fees are set out in 'Engaging an Architects' and a copy can be obtained from RIBA Publications (0207 608 2375) The fees are based upon three classifications of buildings:

Simple – car parks, warehouses, factories etc
Average – offices, retail outlets, housing, schools etc
Complex – specialist buildings, hospitals, laboratories etc

The work stages are set out below

	Stages	Proportion of fee
C	Outline proposals	10% - 15%
D	Detailed proposals	15% - 20%
E	Final proposals	20%
F	Production information	20%
G-L	Tender and construction	25%-35%

The fees are assessed by applying the classification of the project to its value on a fee scale graph. Extra fees may be due if the work is carried out in Stage A (Appraisal) and Stage B (Strategic Brief) or alteration work. These fees would normally be charged on a time basis at an agreed rate. Here are some typical examples of fees for new work.

Value of contract £	Fee as percentage of contract value %
100,000	8.5 – 11.0
500,000	5.8 – 8.5
1,000,000	4.5 – 7.8
2,000,000	4.0 – 7.3
3,000,000	3.8 – 7.0

Quantity Surveyors' fees

The following examples of fees are set out in Professional Charges for Quantity Surveying Services obtainable from Surveyors Publications, (01203 694757) Although Scale 36 (inclusive fees for quantity surveying services) has been abolished it is included here for general guidance.

Scales 36 and 37 – Building work

There are three basic categories of works:

Category A	Complex with little repetition
Category B	Less complex with some repetition
Category C	Simple.

For each category there are two types of fees:

Scale 36	Inclusive scale for complete services
Scale 37	Itemised scale divided into pre- and post-contract services.

Here are some examples of percentages and fees for varying sized projects.

Scale 36 – Inclusive services

Value of work £	Category A £	%	Category B £	%	Category C £	%
150,000	9,380	6.25	9,060	6.04	7,650	5.10
250,000	14,380	5.75	13,760	5.50	11,750	4.70
350,000	19,030	5.44	18,060	5.16	15,450	4.41
450,000	23,330	5.18	21,960	4.88	18,750	4.17
750,000	34,880	4.65	32,010	4.27	27,450	3.66
1,250,000	51,880	4.15	46,010	3.68	39,950	3.20
2,500,000	90,380	3.62	79,010	3.16	68,200	2.73
4,000,000	133,880	3.33	116,010	2.90	99,200	2.48

Scale 37 – Pre-contract services

Value of work £	Category A £	%	Category B £	%	Category C £	%
150,000	4,730	3.15	4,410	2.94	3,930	2.62
250,000	7,030	2.81	6,410	2.56	5,730	2.29
350,000	9,080	2.59	8,160	2.33	7,230	2.27
450,000	10,880	2.42	9,660	2.15	8,430	4.17
750,000	15,830	2.11	13,560	1.81	11,580	1.54
1,250,000	23,330	1.87	19,060	1.52	16,080	1.29
2,500,000	39,080	1.56	31,810	1.27	26,330	1.05
4,000,000	56,380	1.40	45,810	1.15	37,330	0.93

Scale 37 – Post-contract services (Alternative 1)

Value of work £	Category A £	%	Category B £	%	Category C £	%
150,000	3,150	2.10	3,150	2.10	2,520	1.68
250,000	4,850	1.94	4,850	1.94	4,020	1.61
350,000	6,500	1.84	6,450	1.84	5,470	1.56
450,000	8,100	1.80	7,950	1.77	6,870	1.53
750,000	12,450	1.66	11,850	1.58	10,620	1.42
1,250,000	18,950	1.52	17,350	1.52	16,120	1.29
2,500,000	34,200	1.37	30,100	1.20	27,870	1.11
4,000,000	51,200	1.28	44,100	1.10	40,370	1.01

For negotiating and agreeing prices with a contractor:

Value of work £	Fee £	%
150,000	750	0.50
250,000	1,050	0.42
350,000	1,350	0.39
450,000	1,650	0.37
750,000	2,400	0.32
1,250,000	3,350	0.27
2,500,000	4,600	0.18
4,000,000	6,100	0.15

Scale 38 – Civil Engineering work

Category 1   Runways, roads, railways and earthworks and dredging and monolithic walls.

Category 2   Piled quay walls, suspended jetties, bridges, sewers, storage and treatment tanks, turbine halls, reactor blocks.

Pre-contract services

Value of work £	Category 1		Category 2	
	Fee £	%	Fee £	%
500,000	1,960	0.65	3,650	0.73
1,500,000	5,040	0.34	8,850	0.50
2,500.000	7,540	0.30	12,850	0.51
5,000,000	12,790	0.26	21,850	0.44
7,000,000	16,790	0.24	28,850	0.41
12,000,000	25,790	0.21	45,350	0.38
15,000,000	30,290	0.20	54,350	0.36
25,000,000	44,790	0.18	83,350	0.33

Post-contract services

Value of work	Category 1		Category 2	
	Fee	%	Fee	%
£	£		£	
500,000	5,950	1.19	10,750	2.15
750,000	8,250	1.10	15,000	2.00
1,500,000	14,250	0.95	26,250	1.75
2,500.000	20,750	0.83	38,250	1.53
5,000,000	34,000	0.68	65,250	1.31
7,000,000	44,000	0.63	86,250	1.23
12,000,000	68,000	0.57	135,750	1.13
15,000,000	81,500	0.54	162,750	1.09
25,000,000	125,000	0.50	249,750	1.00

Consulting Engineers' fees

Engineers' fees are calculated by dividing the Cost of the Works by the Output Price Index and the resultant figure is then applied to a graph to show the percentage to be set to the final cost of the works. Full details can be obtained from The Association for Consultancy and Engineering (tel. 0207 222 6557).

Landscape Consultants' fees

The fees for work over the value of £10,000 are calculated in two parts. Part 1 is assessed from a graph which indicates the fee percentage from 6% to 14% according to the value of the contract. Part 2 is a coefficient ranging from 1.0 where the consultant has overall responsibility to his client for a job with a normal balance of hard and soft works.

This may increase to 1.2 when the soft works element exceeds 50% of the landscape contract or for private contracts. The coefficient may be decreased to 0.8 for other types of jobs such as golf courses and road landscaping. Here is an example.

Assume a project has a contract value of £100,000 including both hard and soft works in a new business park. The fee graph shows that the percentage norm is 7.5%. The coefficient is 1.2 because the soft work element exceeds 50%. The job coefficient is 1.0 so the compounded coefficient is 1.0 x 1.2. The total percentage fee will be 7.5% x 1.2 = 9.0%.

The scale of fees allows for other methods of remuneration such as lump sum fees, using a ceiling figure in conjunction with a time basis, or having a retainer that can be reviewed after a period and paid according to the value of the actual work carried out.

Similarly, when only occasional work is required this can be charged on a time basis. Site surveys would normally be paid on a lump sum basis of estimated time involved.

Professional team all-in fees

Assessing the fees for professional teams working on a development project can be complicated due to the different methods of fee calculation adopted by the various professional bodies. Worked examples of these are shown in this chapter relating to a range of contract values.

Sometimes the client will prefer to deal with only one discipline and a lead professional will be appointed. This firm will negotiate an overall fee for the whole team and the table below shows the effect of this arrangement. The figures have been rounded off to the nearest £1000.

Project costs (£000)	8% £000	9% £000	10% £000	11% £000	12% £000	13% £000	14% £000	15% £000
200	16	18	20	22	24	26	28	30
300	24	27	30	33	36	39	42	45
400	32	36	40	44	48	52	56	60
500	40	45	50	55	60	65	70	75
600	48	54	60	66	72	78	84	90
700	56	63	70	77	84	91	98	105
800	64	72	80	88	96	104	112	120
900	72	81	90	99	108	117	126	135
1,000	80	90	100	110	120	130	140	150
1,200	96	108	120	132	144	156	168	180
1,400	112	126	140	154	168	182	196	210
1,600	128	144	160	176	192	208	224	240
1,800	144	162	180	198	216	234	252	270
2,000	160	180	200	220	240	260	280	300
2,250	180	202	225	247	270	292	315	337
2,500	200	225	250	275	300	325	350	375
2,750	220	247	275	302	330	357	385	412
3,000	240	270	300	330	360	390	420	450
3,250	260	292	325	357	390	422	455	487
3,500	280	315	350	385	420	455	490	525
3,750	300	337	375	412	450	487	525	562

Project costs (£000)	8% £000	9% £000	10% £000	11% £000	12% £000	13% £000	14% £000	15% £000
4,000	320	360	400	440	480	520	560	600
5,500	440	495	550	605	660	715	770	825
6,000	480	540	600	660	720	780	840	900
7,000	560	630	700	770	840	910	980	1050
8,000	640	720	800	880	960	1040	1120	1200
9,000	720	810	900	990	1080	1170	1260	1350
10,000	800	900	1000	1100	1200	1300	1400	1500

Planning permission fees

These can be quite complex and a summary of some of the main fees are set out below.

Dwelling houses	£220 per 0.1 hectare for outline planning (maximum £5,500 and £220 per house (maximum £11,000).
Alterations and extensions to houses	Single house £110. Two or more houses £220.
Change of use of building for houses	£220 per house (maximum £11,000).

Building regulations

Local authorities are now able to set their own scale of fees for work in connection with building regulations. There are three main charges involved:

Plan charge or *Full Plan* to be paid when plans are submitted.

Building Notice to be paid when the Notice is submitted.

Inspection fee to be paid after first inspection.

Here are some examples of these charges which include VAT.

Small domestic work and alterations

Work	Plan charge £	Building notice £	Inspection fee £
Extension less than 10m2	130	263	133
Extension 10 – 40m2	130	130	nil

8

Useful addresses

Association of Consulting Engineers
Alliance House,
12 Caxton Street
London SW1 OQL
(0207 222 6557)

British Association of Landscape Industries
Landscape House
National Agricultural Centre
Stoneleigh Park
Warwickshire CV8 2LG
(0247 669 0333)

British Board of Agrement
PO Box 195, Bucknall's Lane
Garston, Watford
Herts WD2 7NG
(01923 665300)

British Decorators Association
32 Coton Road
Nuneaton
Warwickshire CV11 5TW
(0247 635 3776)

British Flat Roofing Council
186 Beardall Street
Hucknall
Nottingham NG15 7JU
(0115 956 6666)

British Property Federation
1 Warwick Row
7th Floor
London SW1E 5ER
(0207 828 0111)

British Standards Institution
389 Chiswick High Street
London W4 4AL
(0208 996 9000)

British Woodworking Federation
55 Tufton Street
London SW1 3QL
(0870 458 6939)

Building Centre Group
26 Store Street,
London WC1E 7BT
(0207 692 4000)

Building Cost Information Service
RICS
12 Great George Street
Parliament Square
London SWIP 3AD
(0207 222 7000)

Building Employers Confederation
56-64 Leonard Street
London EC2A 4JX
(0207 608 5000)

Building Research Establishment
Bucknall's Lane,
Garston, Watford
WD2 7JR
(01923 664000)

Chartered Institute of Arbitrators
24 Angle Gate
London EC1V 2RS
(0207 837 4483)

Chartered Institute of Building
Englemere, Kings's Ride,
Ascot, Berkshire 5LS 8BJ
(01344 630700)

Concrete Society
Century House
Telford Avenue
Crowthorne
Berkshire RG45 6YS
(01344 466 007)

Confederation of British Industry
Centre Point,
103 New Oxford Street,
London WC1A 1DU
(0207 379 7400)

Electrical Contractors Association
ESCA House
34 Palace Court,
Bayswater,
London W2 4HY
(0207 313 4800)

Federation of Master Builders
14-15 Great James Street
London WC1N 3DP
(0207 242 7583)

Glass and Glazing Federation
44-48 Borough High Street
London SE1 1XB
(0207 403 7177)

Heating and Ventilation Contractors' Association
ESCA House
34 Palace Court,
London W2 4JG
(0207 313 4900)

Housing Corporation HQ
149 Tottenham Court Road,
London W1P OBN
(0207 393 2000)

Institute of Mechanical Engineers
1 Birdcage Walk,
London SW1H 9JJ
(0207 222 7894)

Institute of Plumbing
64 Station Lane,
Hornchurch,
Essex RN12 6NB
(017108 472791)

Institution of Civil Engineers
1-7 Great George Street,
London SW1P 3AA
(0207 222 7722)

Institution of Civil Engineering Surveyors
Dominion House,
Sibson Road
Sale
Cheshire M33 7PP
(0161 972 3100)

Institution of Electrical Engineers
Savoy Place,
London WC2R OBL
(0207 240 1871)

Institution of Structural Engineers
11 Upper Belgrave Street,
London SW1X 8BH
(0207 235 4535)

Joint Contracts Tribunal
66 Portland Place
London W1N 4AD
(0207 580 5533)

Royal Institute of British Architects
66 Portland Place
London W1N 4AD
(0207 580 5533)

Royal Institute of Chartered Surveyors
12 Great George Street,
London SW1P 3AD
(0207 222 7000)

Royal Town Planning Institute
41 Botolph Lane
London ER3R 8DL
(0207 929 9494)

Timber Research and Development Association
Stocking Lane
Hughenden Valley
High Wycombe
Buckinghamshire
HP14 4ND
(01494 569 600)

Town and Country Planning Association
17 Carlton House Terrace
London SW1Y 5AS
(0207 930 8903)

Water Services Association
1 Queen Ann's Gate,
London SW1H 9BT
(0207 957 4567)

Welsh Development Agency
Treforest Industrial Estate,
Pontypridd,
Glamorgan CS37 5UT
(01345 775577)

Zinc Development Association
6 Wren's Court
56 Victoria Road
Sutton Coldfield
West Midlands
B72 1SY
(0121 355 8386)

9

Estimating data

When preparing first stage estimates, it is often necessary to take a broad view because of lack of time or information. The data in this chapter is intended to help in this process.

LABOUR

The rates in this book have been calculated based on the following labour hourly values:

General operatives	£11
Craftsmen	£15
Plumbers	£16
Electricians	£17

The following hourly gang rates have also been used where applicable.

Building

Groundwork gang	£65
Concreting gang	£50
Steel fixing gang	£60
Formwork gang	£140
Bricklaying gang	£120
Lightweight blockwork gang	£120
Dense blockwork gang	£135
Carpentry gang	£80
Drain laying gang	£25

Civil engineering

Concreting gang	£85
Shuttering gang	£65
Pipe laying (small bore) gang	£60
Pipe laying (large bore) gang	£55
Sub base laying gang	£75
Concrete paving gang	£110

The metric system

Linear

1 centimetre (cm)	=	10 millimetres (mm)
1 decimetre (dm)	=	10 centimetres (cm)
1 metre (m)	=	10 decimetres (dm)
1 kilometre (km)	=	1000 metres (m)

Area

100 sq millimetres	=	1 sq centimetre
100 sq centimetres	=	1 sq decimetre
100 sq decimetres	=	1 sq metre
1000 sq metres	=	1 hectare

Capacity

1 millilitre (ml)	=	1 cubic centimetre (cm3)
1 centilitre (cl)	=	10 millilitres (ml)
1 decilitre (dl)	=	10 centilitres (cl)
1 litre (l)	=	10 decilitres (dl)

Weight

1 centigram (cg)	=	10 milligrams (mg)
1 decigram (dg)	=	10 centigrams (mcg)
1 gram (g)	=	10 decigrams (dg)
1 decagram (dag)	=	10 grams (g)
1 hectogram (hg)	=	10 decagrams (dag)

Conversion equivalents (imperial/metric)

Length

1 inch	=	25.4 mm
1 foot	=	304.8 mm
1 yard	=	914.4 mm
1 yard	=	0.9144 m
1 mile	=	1609.34 m

Area

1 sq inch	=	645.16 sq mm
1 sq ft	=	0.092903 sq m
1 sq yard	=	0.8361 sq m
1 acre	=	4840 sq yards
1 acre	=	2.471 hectares

Liquid

1 lb water	=	0.454 litres
1 pint	=	0.568 litres
1 gallon	=	4.546 litres

Horse-power

1 hp	=	746 watts
1 hp	=	0.746 kW
1 hp	=	33,000 ft.lb/min

Weight

1 lb	=	0.4536 kg
1 cwt	=	50.8 kg
1 ton	=	1016.1 kg

Conversion equivalents (metric/imperial)

Length

1 mm	=	0.03937 inches
1 centimetre	=	0.3937 inches
1 metre	=	1.094 yards
1 metre	=	3.282 ft
1 kilometre	=	0.621373 miles

Area

1 sq millimetre	=	0.00155 sq in
1 sq metre	=	10.764 sq ft
1 sq metre	=	1.196 sq yards
1 acre	=	4046.86 sq m
1 hectare	=	0.404686 acres

Liquid

1 litre	=	2.202 lbs
1 litre	=	1.76 pints
1 litre	=	0.22 gallons

Horse-power

1 watt	=	0.00134 hp
1 kw	=	134 hp
1 hp	=	0759 kg m/s

Weight

1 kg	=	2.205 lbs
1 kg	=	0.01968 cwt
1 kg	=	0.000984 ton

Temperature equivalents

In order to convert Fahrenheit to Celsius deduct 32 and multiply by 5/9.
To convert Celsius to Fahrenheit multiply by 9/5 and add 32.

Fahrenheit	Celsius
230	110.0
220	104.4
210	98.9
200	93.3
190	87.8
180	82.2
170	76.7
160	71.1
150	65.6
140	60.0
130	54.4
120	48.9
110	43.3
100	37.8
90	32.2
80	26.7
70	21.1
60	15.6
50	10.0
40	4.4
30	-1.1
20	-6.7
10	-12.2
0	-17.8

Areas and volumes

Figure	Area	Perimeter
Rectangle	Length × breadth	Sum of sides
Triangle	Base × half of perpendicular height	Sum of sides
Quadrilateral	Sum of areas of contained triangles	Sum of sides
Trapezoidal	Sum of areas of contained triangles	Sum of sides
Trapezium	Half of sum of parallel sides × perpendicular height	Sum of sides
Parallelogram	Base × perpendicular height	Sum of sides
Regular polygon	Half sum of sides × half internal diameter	Sum of sides
Circle	pi × radius²	pi × diameter or pi × 2 × radius

Figure	Surface area	Volume
Cylinder	pi × 2 × radius² × length (curved surface only)	pi × radius² × length
Sphere	pi × diameter²	Diameter3 × 0.5236

Weights of materials	kg/m3	kg/m2	kg/m
Aggregate, coarse	1,500		
Ashes	800		
Ballast	600		
Blockboard, standard	940-1000		
Blockboard, tempered	940-1060		
Blocks, natural aggregate			
75mm		160.00	
100mm		215.00	
140mm		300.00	

Weights of materials	kg/m3	kg/m2	kg/m
Blocks, lighweight aggregate			
75mm		60.00	
100mm		80.00	
140mm		112.00	
Bricks, Fletton		1,820.00	
Bricks, engineering		2,250.00	
Bricks, concrete		1,850.00	
Brickwork, 112.5mm		220.00	
Brickwork, 215mm		465.00	
Brickwork, 327.5mm		710.00	
Cement	1,440		
Chalk	2,240		
Chipboard, standard	650-750		
Chipboard, flooring	680-800		
Clay	1,800		
Concrete	2,450		
Copper pipes, table X			
6mm			0.091
8mm			0.125
10mm			0.158
12mm			0.191
15mm			0.280
18mm			0.385
22mm			0.531
28mm			0.681
35mm			1.133
42mm			1.368
54mm			1.769
Copper pipes, table Y			
6mm			0.117
8mm			0.162
10mm			0.206
12mm			0.251
15mm			0.392
18mm			0.476
22mm			0.697
28mm			0.899
35mm			1.409
42mm			1.700
54mm			2.905

Weights of materials	kg/m3	kg/m2	kg/m
Copper pipes, table Z			
6mm			0.077
8mm			0.105
10mm			0.133
12mm			0.161
15mm			0.203
18mm			0.292
22mm			0.359
28mm			0.459
35mm			0.670
42mm			0.922
54mm			1.334
Flint	2,550		
Gravel	1,750		
Hardcore	1,900		
Hoggin	1,750		
Glass, clear sheet			
3mm		7.50	
4mm		10.00	
5mm		12.50	
6mm		15.00	
10mm		25.00	
12mm		30.00	
15mm		37.50	
19mm		47.50	
25mm		63.50	
Glass, float			
3mm		7.50	
4mm		10.00	
5mm		12.50	
6mm		15.00	
Glass, patterned			
3mm		6.00	
4mm		7.50	
5mm		9.50	
6mm		11.50	
10mm		21.50	
Laminboard	500-700		
Lime, ground	750		

Weights of materials	kg/m3	kg/m2	kg/m
Mild steel flat bars			
25 × 9.53mm			1.910
38 × 9.53mm			2.840
50 × 12.70			5.060
50 × 19.00			7.590
Mild steel round bars			
6mm			0.222
8mm			0.395
10mm			0.616
12mm			0.888
16mm			1.579
20mm			2.466
25mm			3.854
32mm			6.313
40mm			9.864
50mm			15.413
Mild steel square bars			
6mm			0.283
8mm			0.503
10mm			0.784
12mm			1.131
16mm			2.010
20mm			3.139
25mm			4.905
32mm			8.035
40mm			12.554
50mm			19.617
Plaster			
Carlite browning.11mm thick		7.80	
Carlite tough coat, 11mm thick		7.80	
Carlite bonding, 8mm thick		7.10	
Carlite bonding, 11mm thick		9.80	
Thistle hardwall, 11mm thick		8.80	
Thistle dri-coat, 11mm thick		8.30	
Thistle renovating, 11mm thick		8.80	
Sand	1,600		
Screed, 12.5mm thick		29.00	
Stone, Bath	2,200		
Stone, crushed	1,350		
Stone, Darley Dale	2,400		

Weights of materials	kg/m3	kg/m2
Stone, natural	2,400	
Stone, Portland	2,200	
Stone, reconstructed	2,250	
Stone, York	2,400	
Terrazzo, 25mm thick		45.50
Timber		
Ash	800	
Baltic Spruce	480	
Beech	815	
Birch	720	
Box×	960	
Cedar	480	
Ebony	1,215	
Elm	625	
Greenheart	960	
Jarrah	815	
Maple	750	
Pine, Pitchpine	800	
Pine, Red Deal	575	
Pine, Yellow Deal	530	
Sycamore	530	
Teak, African	960	
Teak, Indian	655	
Walnut	495	
Top soil	1,000	
Water	950	
Woodblock flooring		
softwood		12.70
hardwood		17.60
Zinc sheeting		4.6

EXCAVATION AND FILLING

Shrinkage of deposited material

Clay	-10%
Gravel	-7.50%
Sandy soil	-12.50%

Bulking excavated material

Clay	40%
Gravel	25%
Sand	20%

Typical fuel comsumption for plant	Engine size kW	Litres per hour
Compressors up to	20	4.00
	30	6.50
	40	8.20
	50	9.00
	75	16.00
	100	20.00
	125	25.00
	150	30.00
Concrete mixers up to		
	5	1.00
	10	2.40
	15	3.80
	20	5.00
Dumpers	5	1.30
	7	2.00
	10	3.00
	15	4.00
	20	4.90
	30	7.00
	50	12.00
Excavators	10	2.50
	20	4.50
	40	9.00
	60	13.00
	80	17.00
Pumps	5	1.10
	10	2.10
	15	3.20
	20	4.20
	25	5.50

CONCRETE WORK

Concrete mixes	Mix	Cement t	Sand m3	Aggregate m3	Water litres
	1:1:2	0.50	0.45	0.70	208.00
	1:1.5:3	0.37	0.50	0.80	185.00
	1:2:4	0.30	0.54	0.85	175.00
	1:2.5:5	0.25	0.55	0.85	166.00
	1:3:6	0.22	0.55	0.85	160.00

BRICKWORK AND BLOCKWORK

Bricks per m2 (brick size 215 × 103.5 × 65mm)

Half brick wall
  stretcher bond   59
  English bond   89
  English garden wall bond   74
  Flemish bond   79

One brick wall
  English bond   118
  Flemish bond   118

One and a half brick wall
  English bond   178
  Flemish bond   178

Two brick wall
  English bond   238
  Flemish bond   238

Metric modular bricks

  200 × 100 × 75mm
    90mm thick   133
    190mm thick   200

200 × 100 × 100mm

90mm thick	50
190mm thick	100
290mm thick	150

300 × 100 × 75mm

90mm thick	44

300 × 100 × 100mm

90mm thick	50

Blocks per m2 (block size 414 × 215mm)

60mm thick	9.9
75mm thick	9.9
100mm thick	9.9
140mm thick	9.9
190mm thick	9.9
215mm thick	9.9

Mortar per m2	**Wirecut m3**	**1 Frog m3**	**2 Frogs m3**
Brick size 215 × 103.5 × 65mm			
Half brick wall	0.017	0.024	0.031
One brick wall	0.045	0.059	0.073
One and a half brick wall	0.072	0.093	0.114
Two brick wall	0.101	0.128	0.155

Brick size 200 × 100 × 75mm	**Solid m3**	**Perforated m3**
90mm thick	0.016	0.019
190mm thick	0.042	0.048
290mm thick	0.068	0.078

Brick size 200 × 100 × 100mm	**Solid m3**	**Perforated m3**
90mm thick	0.013	0.016
190mm thick	0.036	0.041
290mm thick	0.059	0.067

	Solid m3	Perforated m3
Brick size 200 × 100 × 100mm		
90mm thick	0.015	0.018
Block size 440 × 215mm		
60mm thick	0.004	
75mm thick	0.005	
100mm thick	0.006	
140mm thick	0.007	
190mm thick	0.008	
215mm thick	0.009	

MASONRY

Mortar per m2 of random walling	m3
300mm thick wall	0.120
450mm thick wall	0.160
550mm thick wall	0.120

CARPENTRY AND JOINERY

Length of boarding required	m/m2
Board width, 75mm	13.33
Board width, 100mm	10.00
Board width, 125mm	8.00
Board width, 150mm	6.67
Board width, 175mm	5.71
Board width, 200mm	5.00

ROOFING	Lap mm	Gauge mm	Nr/m2	Battens m/m2
Clay/concrete tiles				
267 × 165mm	65	100	60.00	10.00
	65	98	64.00	10.50
	65	90	68.00	11.30

	Lap mm	Gauge mm	Nr/m2	Battens m/m2
387 × 230mm	75	300	16.00	3.20
	100	280	17.40	3.50
420 × 330mm	75	340	10.00	2.90
	100	320	10.74	3.10
Fibre slates				
500 × 250mm	90	205	19.50	4.85
	80	210	19.10	4.76
	70	215	18.60	4.65
600 × 300mm	105	250	13.60	4.04
	100	250	13.40	4.00
	90	255	13.10	3.92
	80	260	12.90	3.85
	70	263	12.70	3.77
400 × 200mm	70	165	30.00	6.06
	75	162	30.90	6.17
	90	155	32.30	6.45
500 × 250mm	70	215	18.60	4.65
	75	212	18.90	4.72
	90	205	19.50	4.88
	100	200	20.00	5.00
	110	195	20.50	5.13
600 × 300mm	100	250	13.4	4.00
	110	245	13.60	4.08
Natural slates				
405 × 205mm	75	165	29.59	8.70
405 × 255mm	75	165	23.75	6.06
405 × 305mm	75	165	19.00	5.00
460 × 230mm	75	195	23.00	6.00
460 × 255mm	75	195	20.37	5.20
460 × 305mm	75	195	17.00	5.00
510 × 255mm	75	220	18.02	4.60
510 × 305mm	75	220	15.00	4.00

	Lap mm	Gauge mm	Nr/m2	Battens m/m2
560 × 280mm	75	240	14.81	4.12
560 × 280mm	75	240	14.00	4.00
610 × 305mm	75	265	12.27	3.74

Reconstructed stone slates

	Lap mm	Gauge mm	Nr/m2	Battens m/m2
380 × 250mm	75	150	16.00	3.20

PLASTERING AND TILING

Plaster coverage

	m2 per 1000kg
Carlite browning, 11mm thick	135-155
Carlite tough coat, 11mm thick	135-150
Carlite bonding, 11mm thick	100-115
Thistle hardwall, 11mm thick	115-130
Thistle dri-coat, 11mm thick	135-135
Thistle renovating, 11mm thick	115-125

Tile coverage

	Nr
152 × 152mm	43.27
200 × 200mm	25.00

PLUMBING AND HEATING

Roof drainage

	Area m2	Pipe mm	Gutter mm
One end outlet	15	50	75
	38	68	100
	100	110	150
Centre outlet	30	50	75
	75	68	100
	200	110	150

PAINTING AND WALLPAPERING

Average coverage of paints m2 per litre	Timber	Plastered surfaces	Brickwork
Primer	10-12	9-11	5-7
Undercoat	10-12	11-14	6-8
Gloss	11-14	11-14	6-8
Emulsion	10-12	12-15	6-10

Wallpaper coverage per roll	Rolls nr	Wall height m	Room perimeter m
	4	2.50	8
	5	2.50	9
	5	2.50	10
	6	2.50	11
	6	2.50	12
	7	2.50	13
	7	2.50	14
	8	2.50	15
	8	2.50	16
	8	2.50	17
	9	2.50	18
	10	2.50	19
	10	2.50	20
	10	2.50	21
	11	2.50	22
	11	2.50	23
	12	2.50	24
	13	2.50	25
	13	2.50	26
	14	2.50	27
	5	2.80	8
	5	2.80	9
	5	2.80	10
	7	2.80	11
	7	2.80	12
	7	2.80	13
	8	2.80	14
	8	2.80	15

Rolls nr	Wall height m	Room perimeter m
9	2.8	16
10	2.8	17
10	2.8	18
11	2.8	19
11	2.8	20
12	2.8	21
13	2.8	22
13	2.8	23
14	2.8	24
14	2.8	25
15	2.8	26
15	2.8	27

Drainage

Trench widths

	Under 1.5m deep mm	Over 1.5m deep mm
Pipe diameter 100mm	450	600
Pipe diameter 150mm	500	650
Pipe diameter 225mm	600	750
Pipe diameter 300mm	650	800

**Volumes of filling for pipe
bed and haunching
(m3 per m)**

	m3
Pipe diameter 100mm	0.117
Pipe diameter 150mm	0.152
Pipe diameter 225mm	0.195

**Volumes of filling for pipe
bed and surround
(m3 per m)**

Pipe diameter 100mm	0.185
Pipe diameter 150mm	0.231
Pipe diameter 225mm	0.285
Pipe diameter 300mm	0.391

LANDSCAPING

Tree sizes	Type	Height	Clear stem height	Girth
	Light standard	2.50-2.75m	1.50-1.80m	6-8cm
	Standard	2.75-3.00m	1.8m	8-10cm
	Selected standard	3.00-3.50m	1.8m	10-12cm
	Heavy standard	3.50-4.00m	1.8m	12-14cm
	Extra heavy standard	4.00-5.00m	1.8m	14-16cm

Quantities of seed for sports fields

	gm/m2 34	gm/m2 50	gm/m2 102	gm/m2 500
Bowling green (size 38.4 × 38.4m, area 1,475m2)	50	75	150	200
Cricket square (size 22.8 × 22.8,. area 522m2)	18	27	54	72
Golf green (18 nr each size 570m2)	350	525	1050	1400
Lawn tennis (size 36.6 × 18.3m, area 670m2)	23	35	69	92
Football (size 119 × 91m, area 10,380m2)	368	552	1104	1472
Rugby (size 100 × 69m, area 6,900m2)	235	352	705	940

	34gm per m2	50gm per m2	102gm per m2	500gm per m2
Hockey (size 91 × 55m, area 5,005m2)	170	255	510	680

Weedkiller

	litres per hectare
Bowling green (size 38.4 × 38.4m, area 1,475m2)	1.48
Cricket square (size 22.8 × 22.8m, area 522m2)	0.52
Golf green (18 nr each size 570m2)	10.28
Lawn tennis (size 36.6 × 18.3m, area 670m2)	0.67
Football (size 119 × 91m, area 10,380m2)	10.83
Rugby (size 100 × 69,. area 6,900m2)	6.90
Hockey (size 91 × 55m, area 5,005m2)	5.00

CIVIL ENGINEERING

Outputs

Placing ready-mixed concrete	Labour gang (m3 per hour)
Mass concrete	
blinding	
150mm thick	5.50
150-300mm thick	6.25
300-500mm thick	7.00
bases	
not exceeding 150mm thick	5.00
not exceeding 300mm thick	5.75
not exceeding 500mm thick	6.75
exceeding 150mm thick	7.00
Reinforced concrete	
bases	
not exceeding 300mm thick	5.50
not exceeding 500mm thick	6.25
exceeding 150mm thick	6.75
suspended slabs	
not exceeding 150mm thick	3.75
not exceeding 300mm thick	4.75
exceeding 300mm thick	5.75
walls	
not exceeding 150mm thick	3.50
not exceeding 300mm thick	4.50
exceeding 300mm thick	5.00
beams and columns	
sectional area not exceeding 0.03m2	2.00
sectional area not exceeding 0.03 -1.0m2	2.50
sectional area exceeding 1.0m2	3.50

Fixing bar reinforcement

The following figures exclude delivery, craneage and hoisting in position. The hours refer to tonnes per hour for a steelfixer and general operative.

	Height up to 6m (hours)	Height 7 to 12m (hours)	Height 13 to 19m (hours)	Height over 19m (hours)
Straight round bars				
to beams, walls, roofs and walls	0.06	0.08	0.12	0.16
to braces, columns and sloping roofs	0.06	0.08	0.12	0.16
Bent round bars				
to beams, walls, roofs and walls	0.04	0.06	0.06	0.08
to braces, columns and sloping roofs	0.02	0.02	0.04	0.06
Straight, indented or square bars				
to beams, walls, roofs and walls	0.04	0.08	0.10	0.14
to braces, columns and sloping roofs	0.02	0.04	0.04	0.08
Bent, indented or square bars				
to beams, walls, roofs and walls	0.04	0.04	0.06	0.06
to braces, columns and sloping roofs	0.02	0.02	0.04	0.04

Erecting and striking formwork	**Joiner** $m2$ per hour	**General operative** $m2$ per hour
Vertical wall face, height		
up to 1.5m	1.70	0.80
1.5 to 3.0m	1.40	0.70
3.0 to 4.5m	1.20	0.60
4.5 to 6.0m	1.00	0.50
Horizontal slabs, height		
up to 3.0m	1.11	1.11
3.0 to 3.6m	1.05	1.05
3.6 to 4.2m	1.00	1.00
4.2 to 4.8m	0.95	0.95
4.8 to 5.4m	0.90	0.90
5.4 to 6.0m	0.83	0.83

Formwork multipliers

	Multiplier
Battered walls	1.20
Circular walls to large radius	1.70
Circular walls to small radius	2.10
Formwork, one use	1.00
Formwork, two uses, per use	0.85
Formwork, three uses, per use	0.75
Formwork, four uses, per use	0.72
Formwork, five uses, per use	0.68
Formwork, six uses, per use	0.66
Formwork, seven uses, per use	0.63

Drainage

	Unit	Drainage gang (m/hour) in trenches less than 1.5m deep	Drainage gang (m/hour) in trenches less than 3.0m deep	Drainage gang (m/hour) in trenches less than 4.5m deep
Lay and joint flexible-jointed clayware pipes, diameter				
100mm	m	10	8	7
150mm	m	7	6	5
225mm	m	5	4	3
300mm	m	4	3	2
375mm	m	3	2	2
450mm	m	2	2	1
Bends				
100mm	nr	20	17	14
150mm	nr	17	14	13
225mm	nr	13	10	8
300mm	nr	10	8	7
375mm	nr	7	6	5
450mm	nr	3	3	2
Single junctions				
100mm	nr	13	10	8
150mm	nr	7	6	5
225mm	nr	6	5	4
300mm	nr	4	3	3
375mm	nr	3	2	2
450mm	nr	2	2	1

Precast concrete manholes	Unit	Pipelayer	General operative
675mm diameter shaft rings	m/hour	1.00	0.30
900mm diameter manhole rings	m/hour	0.60	0.20
1200mm diameter manhole rings	m/hour	0.40	0.10
1500mm diameter manhole rings	m/hour	0.30	0.10
900 to 675mm tapers	nr/hour	1.00	0.30
1200 to 675mm tapers	nr/hour	0.60	0.20
900 to 675mm tapers	nr/hour	0.50	0.20
Cover slabs to 675mm rings	nr/hour	2.50	0.80
Cover slabs to 675mm rings	nr/hour	2.00	0.70
Cover slabs to 675mm rings	nr/hour	1.40	0.50
Cover slabs to 675mm rings	nr/hour	0.90	0.30
Build in pipes and make good			
150mm diameter	nr/hour	4.00	0.00
300mm diameter	nr/hour	2.00	0.00
450mm diameter	nr/hour	1.00	0.00
Benching 150mm thick	m2/hour	1.20	1.20
Benching 300mm thick	m2/hour	0.60	0.60
Render benching 25mm thick	m2/hour	1.10	1.10
Manhole cover and frame	nr/hour	1.30	1.30

Index